To: Arthur & Elaine,
with my best wishes,
and cordial
greetings!

*[signature]*

February 27, 2007.

# *biological* wealth
## & other essays

# By the Same Author

Haldane and Modern Biology
Haldane, The Life and Work of J. B. S. Haldane with Special Reference
   to India
Cleft Lip and Palate, Aspects of Reproductive Biology
The Foundations of Human Genetics
Selected Genetic Papers of J. B. S. Haldane
The History and Development of Human Genetics
If I am To Be Remembered, The Life and Work of Julian Huxley
Haldane's Daedalus Revisited
Science and Society, An Indo-American Perspective
Biological and Social Issues in Biotechnology Sharing
Haldane in India
Profiles in Genetics
Evolutionary Aspects of Infectious Disease (in press)

# biological wealth & other essays

## K. R. Dronamraju

*President, Foundation for Genetic Research*
*Houston, Texas, USA.*

**World Scientific**
*New Jersey • London • Singapore • Hong Kong*

*Published by*

World Scientific Publishing Co. Pte. Ltd.

P O Box 128, Farrer Road, Singapore 912805

*USA office:* Suite 1B, 1060 Main Street, River Edge, NJ 07661

*UK office:* 57 Shelton Street, Covent Garden, London WC2H 9HE

**British Library Cataloguing-in-Publication Data**
A catalogue record for this book is available from the British Library.

**BIOLOGICAL WEALTH AND OTHER ESSAYS**

Printed in Singapore by World Scientific Printers (S) Pte Ltd

# *Dedication*

I am pleased to dedicate this book of essays to Jim Watson, who has contributed profoundly to revolutionizing biology. Not since Darwin have we witnessed such a fundamental transformation of biological sciences. DNA is a household word today. However, as Jim wrote, "Unfortunately, lots of well-intentioned outsiders today see recombinant DNA as a test case for the scientist's responsibility to society" (from *A Passion for DNA: Genes, Genomes and Society*, p. 57).

These essays reflect a similar concern about a much misunderstood opposition to DNA research, on the one hand, and, paradoxically, its commercialization on the other.

# *Foreword*

This book on biological wealth by Prof. Krishna R. Dronamraju is a timely publication. Research on genetic modification with particular reference to crop plants is now at a crossroad. On the one hand, recombinant DNA technology has opened up uncommon opportunities for developing novel genetic combinations. Some of them can be of immense value in the battle against biotic and abiotic stresses. Others may be of help in bringing about a nutrition revolution. On the other hand, there is serious apprehension in the public mind about the food and environmental safety aspects of genetically modified crops. Ethical considerations are also becoming important not only in technology choice but also in the equitable sharing of benefits.

Biodiversity is the feed stock of the biotechnology industry. Unfortunately, the primary conservers of economically valuable biodiversity and the holders of traditional knowledge tend to remain poor, in contrast to the prosperity of those using their knowledge and material. This is why the legally binding Convention on Biodiversity places great stress on the ethical and equity aspects of benefit sharing. The Cartegena Protocol on Biosafety provides an international framework for dealing with the biosafety aspects of transgenic plants and animals.

Prof. Dronamraju has dealt with all these issues in a lucid and authoritative manner. This book shows the way of ensuring that the benefits of the gene revolution triggered by recombinant DNA technology reach the unreached. It also emphasizes the need for a strong ethical push to match the technological pull represented by genomics, proteomics, the internet and genetic engineering. Hence, this book should be widely read by all concerned with the use of genetic modification technologies for fostering sustainable food and health security.

M. S. Swaminathan

# *Introduction*

Several of these essays were written for a semi-popular audience in three countries: the United States, India and Great Britain during the years 1997–2001. They cover a number of topics which have arisen since the successful application of DNA technology. They include various issues and controversies that are certainly topical today, such as biodiversity, genetic engineering, genetically modified organisms, the genome project, intellectual property rights, cloning, stem cell research, GATT, WTO, CBD, and so on. Biological research, in particular the applied aspect of biology that we call biotechnology, has changed radically over the last twenty years. Much of it is driven by commercial and market forces, which are bent on exploiting every new discovery and observation for financial gain.

These developments have led to some interesting consequences of global importance. Initially, the great discovery of the molecular structure of DNA by James Watson and Francis Crick in 1953 ushered in the DNA revolution, which is clearly the most important event in biology since the Darwinian revolution of the nineteenth century. Biology used to be regarded as a rather unglamorous subject, whereas the physical sciences, especially nuclear physics (and later particle physics), occupied a much loftier place in universities as well as in the public mind. The development of nuclear weapons for security purposes was partly responsible for that state of affairs. However, the tide turned rapidly in favor of biology with the discovery of recombinant DNA in 1973 by Herbert Boyer and Stanley Cohen. It was already clear by then that the new technology had the potential to revolutionize both agriculture and medicine in a profound way. Indeed, the prospect of recombinant DNA contaminating everything in the world was considered so dangerous that a conference of distinguished molecular biologists was convened shortly afterwards at Asilomar in California to assess the potential risks of any future experiments involving recombinant DNA. Almost thirty years later, many of those concerns, fortunately, remain groundless. For instance, in reviewing a large number of protocols for somatic cell gene therapy in humans, we found the experiments to be no more dangerous than many others in biology and perhaps less so than some. I was a member of the Recombinant DNA Advisory Committee (RAC) of the U.S. National Institutes of Health from 1992–1996. The unfortunate death of one patient undergoing gene therapy in Philadelphia was due to the sloppy work of the investigators.

## Biological wealth

The DNA revolution in recent decades has focused attention on the biological wealth of our planet that is exemplified in the biodiversity that is around us. As the economic and scientific implications of the biological wealth of our planet have become more obvious in recent years, commercial and business interests have taken a keen interest in exploiting these new opportunities. Commercial exploitation, in turn, has led to a vast number of intellectual property claims to biological materials that were unheard of twenty years ago. The first patent issued for a "man-made" living organism was in relation to the bacterium, *Pseudomonas aeruginosa*, in 1981, in the celebrated case of Diamond versus Chakrabarty. These developments have greatly blurred the boundaries between biological research and commercial or industrial application.

The commercial success of biotechnology has led to an explosive growth of small private companies, attempting to develop various products of importance to agriculture and medicine. These, in turn, led to an extensive intellectual property rights (IPR) system which was primarily developed by the United States and western Europe. It is now widely recognized all over the world that a sound IPR is a prerequisite for global trade and especially the transfer or sharing of technology. However, this problem opened a "Pandora's Box," leading to claims and counterclaims and even abuse or misuse of the system in several countries. False claims and non-recognition of indigenous contributions abound. Industrialized countries have had a clear advantage at the outset. Many people in developing countries have come to believe that the IPR system is a new variant of economic imperialism where the developed countries dictate the "rules" of the "game." In the meantime, the economic programs of agencies such as the GATT, the WTO, the IMF and the World Bank have tended to promote the world order that was established after the second world war where developing countries contribute to the economies of developed countries by providing cheap labor and natural resources including biodiversity in perpetuity.

It is ironic that the perpetuation of the new world order is administered and promoted by the educational elite of the developing countries who are largely educated in western Europe and the United States and are imbued with a sense of responsibility to carry on that mission. They have come to play a key role in perpetuating an international economic order in which rich countries continue to become richer and poor nations have clearly become poorer. This situation is not unlike the indentured labor that existed during colonial times when a few small European countries were able to enslave entire sections of Asia, Africa, Central and South America; the only difference being that now it is the developing countries that are indentured in a system that is clearly exploitative. There are powerful forces which are employed for the sole purpose of maintaining the *status quo*. Even in peace time, huge armies, expensive weapons and intelligence networks are maintained at great cost by developed countries. The slightest deviation by a developing country from the established economic order can be met not only with economic and political retaliation, but the threat of armed invasion as well.

Such international agencies as the World Trade Organization (WTO), the World Bank and the International Monetary Fund (IMF) are dominated by the large multinationals from the United States and Europe whose economic interests they serve. These rich and powerful businesses and industries of the private sector exercise their influence outside the democratic framework of the family of nations. No country or community elected the officials of these companies through a democratic process, yet their decisions and policies shape the destinies of billions of people on a global scale.

These facts are relevant to the future of our biological wealth or biodiversity because much of world's biodiversity is located in the poorest countries. Economic policies, industrialization and exploitation of biodiversity adversely impact on the biosphere in several ways. It is easy to see the consequences that have resulted from the post-second world war economic order. Seventy percent of the world's $CO_2$ emissions and the highest proportion of CFCs are emitted by the industrialized nations which contain only about twenty-three percent of the world's human population. On the other hand, developing countries with seventy-seven percent of world's population account for only thirty percent of the world's emissions. In *per capita* terms, the emission of carbon is ten times higher in industrial countries than in developing countries. The United States emits about twenty-five percent of all carbon emissions while consuming about twenty-five percent of all petroleum produced, even though its population is only about five percent of the world's population.

Extensive and repeated expeditions for rare medicinal plant species by those engaged in pharmaceutical development, both in the private sector as well as in the public sector (e.g. The U.S. National Cancer Institute), are resulting in the extinction of valuable plant species. Environmental pollution, human population growth and commercial exploitation have reduced the numbers of plant and animal species drastically.

According to the 2000 Red List published by the International Union for the Conservation of Nature (IUCN), a total of 11,046 species of plants and animals are threatened, facing a high risk of extinction in the near future. This includes twenty-four percent of mammal species and twelve percent of bird species. Most threatened mammals and birds are found in Indonesia, India, Brazil and China, whereas plant species are declining rapidly in South and Central America, Central and West Africa, and Southeast Asia. In the last 500 years, human activity has forced 816 species to extinction, but the rate of extinction in recent years has increased very greatly. Many species are lost even before they are discovered.

Approximately, twenty-five percent of reptiles, twenty percent of amphibians and thirty percent of fish are listed as threatened by the IUCN. A total of 5611 threatened plant species were listed by the IUCN in 2000, but as only four percent of the world's total plant species have been evaluated, the true percentage of threatened plant species could be much higher.

## Economic inequality

According to a report issued by the United Nations Development Programme (UNDP), over one billion people lack even the most basic consumption requirements although the total global consumption of goods and services exceeds US$25 trillion per year. Of the 4.4 billion people who live in developing countries, almost three-fifths lack basic sanitation, one-third safe drinking water, one-quarter adequate housing, and one-fifth access to a modern health service or primary education.

The following statistics are of interest:

| Category | Wealthiest countries | Poorest countries |
| --- | --- | --- |
| Meat and fish | 45% of global supply | less than 5% |
| Protein intake | | |
| (average/day) | 115 grammes | 32 grammes |
| Energy consumption | 58% of global output | 4% |
| Telephone lines | 74% | 1.5% |
| Paper consumption | 84% | 1.1% |
| Vehicles | 87% | less than 1% |

Source: United Nations Development Programme (UNDP), Annual Report 1999.

Intellectual property rights and biodiversity are playing an increasingly important role in the world economy. As mentioned earlier, much of the global biodiversity resides in the developing countries. Yet, these countries lack sufficient resources and trained personnel to enjoy the full economic benefits of their own biodiversity. On the other hand, the developed countries have been able to reap the full benefits of the biodiversity that is owned by the developing countries. Indigenous communities, who have traditionally conserved and maintained bio-diversity in their areas, rarely receive a share of these profits.

To understand the economic disparity between the rich and poor countries, consider the following facts: (a) Half the world — nearly three billion people — live on less than two U.S. dollars a day, (b) The GDP (Gross Domestic Product) of the world's poorest 48 nations (i.e. a quarter of the world's countries) is less than the wealth of the world's three richest people combined, (c) nearly a billion people entered the twenty-first century unable to read or write, and (d) less than one percent of what the world spends every year on weapons would be needed to put every child into school (Source: Global Economic Prospects, World Bank Report 2001).

Most of the increase in debt owed by the poor, developing countries during the 1990s was to pay interest on existing loans. During the years 1990–1997, developing countries paid more to service existing debts than they received in new loans, a total transfer of US$77 billion from the poor countries to the rich. In other words, this is the equivalent of a "global ghetto," a poverty cycle from which the developing countries could never hope to get out in the foreseeable future.

The World Bank's 2001 analysis of long-term economic trends shows how the economic disparity between the richest and the poorest countries has widened (Source: Globalization, Growth and Poverty, World Bank Report 2001):

3 to 1 in 1820
11 to 1 in 1913
35 to 1 in 1950
44 to 1 in 1973
72 to 1 in 1992

Those who are opposed to the WTO, the IMF and the World Bank argue that these organizations are designed to perpetuate and widen the economic gap between the rich and poor nations.

"Globalization" itself has become the latest "dirty" word, a symbol of economic oppression and a tool that aids the rich nations in oppressing the poor. Multinational corporations are said to be the chief beneficiaries of "globalization."

## Social and international issues

The topics discussed in this book should interest all educated readers, although I do not expect them to agree with all my views. They should also interest students of sociology and international relations dealing with issues in technology transfer or sharing. These essays reflect the complex issues that are generated by various individual, institutional, national and global issues involving intellectual property rights and biodiversity.

There are several international issues regarding the ownership of plant and animal species. These are reflected in the statements issued by the Convention on Biological Diversity (CBD), the World Trade Organization (WTO) and other U.N.-associated organizations. National sovereignty is often transgressed by claims of international patents, such as those filed for *basmati* rice and the medicinal derivatives of the *neem* tree, both of which have been used by the people of India for thousands of years.

These and other related issues bearing on the economic benefits of biodiversity, the growing gap in biotechnology between the rich and poor nations as well as other aspects of science in developing countries are discussed in these essays.

## Acknowledgements

In developing these ideas and concepts over the last several years, I have derived much intellectual stimulation from contacts with many colleagues and friends, especially Prof. M. S. Swaminathan, Prof. Joshua Lederberg, Sir Arthur C. Clarke, and, above all, my mentor J. B. S. Haldane.

I am much indebted to the massive compendium, *Cultural and Spiritual Values of Biodiversity*, which was edited by Darrel Addison Posey of Brazil, the Oxford Centre for the Environment, Ethics and Society in the United Kingdom, and several co-editors, and published by Intermediate Technology Publications, London (U.K.), under the auspices of the United Nations Environment Programme (UNEP) in 1999. In particular, I have benefitted greatly by reading the chapter on *"Ethnoscience, TEK and its application to conservation"* by L. Jan Slikkerver, the chapter entitled *"Forests, culture and conservation"* by

Sarah A. Laird, and another on *"Ethical, moral and religious concerns"* by Jeff Golliher. Furthermore, other writings of M. S. Swaminathan, Klaus Topfer, Vernon Heywood and Madhav Gadgil have been most helpful. I would like to thank Michele Wambaugh for assistance in preparing the manuscript for publication.

Krishna R. Dronamraju
*President, Foundation for Genetic Research*
*Houston, Texas, U.S.A.*

# Contents

# BIOLOGICAL WEALTH AND INTELLECTUAL PROPERTY ISSUES

# Rights Over Biological Wealth

Recent years have seen the emergence of a host of controversial issues related to sharing and transfer of technologies. It is Intellectual Property Rights (IPR) that occupy a central place in the ongoing negotiations. The issue of IPR with special reference to biodiversity and plant genetic resources from the viewpoint of the developing world is discussed below.

## Constraints on the acceptance of new technologies

Due to several historic, cultural and economic reasons, developing countries have been slow to accept new developments in biotechnology. For instance: lack of legislation to adopt a sophisticated system of IPR, lack of trained leaders and technical manpower in new technologies, lack of regulatory agencies to monitor or safeguard against possible toxic effects, sensitivity in some countries to patenting living cells, tissues and microorganisms, lack of compensation for indigenous practices and traditional knowledge, fear of dependency on industrialize countries, economic exploitation, loss of raw materials, biodiversity concerns, biological and chemical contamination, absence of public education, lack of awareness and political support for biotechnology transfer, and finally, a host of complex socio-political factors and infrastructure problems which ultimately determine the acceptance of biotechnology.

## Convention on biological diversity

The Convention on Biological Diversity (CBD) was negotiated before the United Nations Conference on Environment and Development (UNCED), held in Rio de Janeiro in 1992. The first meeting of the conference was held in the Bahamas in 1994.

The objectives of the CBD included conservation of biological diversity, sustainable use of its components and equitable share of benefits arising out of the utilization of genetic resources, including transfer of relevant technologies.

Biodiversity is an index of the biological wealth of this planet. The Global Biodiversity Convention became effective in December, 1993. In part, it states that countries have sovereign rights over their own biological resources and are responsible for conserving and using their biological diversity and resources in a sustainable manner.

However, commercial exploitation of biodiversity continues to take place, especially in developing countries, and has led to the rapid extinction of many species.

Following the Uruguay Round of negotiations, developing countries have realized that they must adopt laws and regulations pertaining to intellectual property rights, a notion that is essentially conceived in industrialized countries. Traditional property systems and conventional practices are being replaced by laws of international agreements, which are designed in the industrialized nations.

This has resulted in much apprehension and economic uncertainty with regard to the future status of international trade as well as the continued enjoyment of the local bio-resources of developing nations.

The sovereignty of their local biological resources is also no longer guaranteed. Traditional intellectual properties of the developing nations, such as the farming practices which have conserved crop germplasm for centuries, and medicinal application of various plant derivatives, are neither recognized nor included in the intellectual property laws designed in the West. Western multinational companies are filing patents to stake claim on processes and products which have already been well-recognized in the developing world for centuries, the only difference being that the western products may have been produced by different methods.

Commercial exploitation of the intellectual and physical property of the developing world involves a lopsided process which invariably results in the flow of bio-resources from the developing world to the industrialized nations. Moreover, the economic benefits that result from this process are largely accrued by the industrialized world, with the developing nations receiving only a small fraction of the pie.

## Plant genetic resources

The impact of Intellectual Property Rights is obvious in the whole range of issues related to plant genetic and crop resources. The World Resources Institute (WRI) has termed biochemical and genetic resources the "oil of the information age." Among the issues which are being debated are the North-South dialogue on sharing the world's natural resources, ownership of intellectual property and compensation for indigenous practices as well as farmers' rights.

These issues are being debated in several international conferences aimed at reducing border barriers and restrictions under the auspices of the General Agreement on Tariffs and Trade (GATT). For the first time in GATT negotiations, the Uruguay Round was marked by an active participation by developing countries.

One important development was the stipulation that all signatories should adopt Trade-Related Intellectual Property Rights Protection (TRIPS). Another development was the liberalization of certain conditions for investment in developing countries by domestic and foreign firms, such as exports and export targets. It was further agreed that trade in services would be brought under multilateral trade disciplines in the World Trade Organization (WTO). Exploitation of the biodiversity resources of developing nations by individuals, groups or multinational companies is not often accompanied by conservation programs.

Technology places more power in the hands of a few individuals in the industrialized countries when compared to the great majority of humanity. The immediate effect of implementing Intellectual Property Rights (IPR) today is to widen the gap between rich and poor nations.

Poor countries pay more for certain goods, some developed from naturally occurring species in their own backyard, while industrialized nations benefit.

## Farmers' rights

In recent years, agronomist Dr. M. S. Swaminathan[1] eloquently advocated the concept of plant breeders and farmers' rights in India. His work may serve as a blueprint for several developing nations.

The aims of the proposed act of legislation, entitled "Plant Variety Recognition and Rights Act," include "conservation, evaluation and sustainable utilization of plant genetic resources, to revive and strengthen the *in situ* conservation of land races and folk varieties, which are the result of a thousand years of natural and human selection," and a recognition of the value of indigenous practices and traditional knowledge in terms of intellectual property rights, and so on.

Swaminathan emphasized the important difference between developing and developed nations in this regard: "Many people who discuss this matter do not appreciate the differences between us [India] and industrialized countries where farming is today really agribusiness. Each farm may be about 1,000 hectares. Here, it is one hectare or below in the case of most farms; hence, there is a need for achieving a proper match between legislation and real life conditions." Legislation protecting agrobiodiversity in developing countries, which includes farmers' and plant breeders' rights, is an important aspect of future global food security because agrobiodiversity is more concentrated in the tropics and sub-tropics.

Legislation concerning plant breeders' rights (PBR) has recently been introduced in five Latin American countries, including Argentina, Chile, Colombia, Mexico and Uruguay. One of the main reasons for introducing PBR in Latin America was to improve the transfer of foreign plant breeding technology to domestic plant breeders, seed propagators and growers by assuring foreign breeders that their varieties would enjoy legal protection against unauthorized use. This subject has been reviewed by van Wijk and Jaffe.[2]

## Medicinal plants

The IUCN (International Union for the Conservation of Nature) and the WWF (World Wildlife Fund) estimate that 60,000 higher plant species could become extinct by the middle of the next century unless adequate conservation measures are implemented to reverse the present trend. The threat of extinction is even more realistic when considering the exploitation of medicinal plants, the majority of which are grown in developing countries. The world demand for pharmaceuticals has steadily been going up. Khalil[3] reviewed the problem of IPR with respect to the conservation of medicinal plants in the developing world. Many forest communities have traditionally conserved these species and preserved the knowledge concerning their medicinal properties, yet they are not protected by rights of the kind conceived in the industrialized nations. Frequent screenings of medicinal plant species by foreign expeditionary forces result in the extinction of several species. It is not widely known whether this threat of extinction is even more realistic when considering the exploitation of medicinal plants, the majority of which are grown in developing countries. Neither is it widely known whether such extinction adversely affects the primary healthcare system of the indigenous communities for whom the species are sometimes the only remedy available. Among several instances cited by Khalil, one involved plants found in Kenya, <u>Maytenus Buchanii</u>, which has traditionally been used by the indigenous Digo community for treating cancer. The U.S. National Cancer Institute (N.C.I.O.) took the whole stock for research purposes while the Digo community's rights were not recognized. Additional instances can be cited from several other countries including Ethiopia, Madagascar and Jamaica. The major objective of the plant exploration program, sponsored by the National Cancer Institute, is to collect plants for anti-cancer and anti-HIV evaluation. The countries covered in Asia have included Indonesia, the Philippines, Malaysia, Papua New Guinea, Thailand, Taiwan, Nepal, Pakistan and China. Mays *et al.*[4] has examined the problem of equitable compensation in relation to pharmaceutical production involving biological materials from developing countries. Although the National Cancer Institute has utilized the legal concept of *quid pro quo* in designing its "Letter of Collection" (LOC) by the NCI, the problem of sustained conservation of the species in the source countries has not been addressed. On the other hand, in Costa Rica, biodiversity conservation is a central feature of the INBio-Merck[*] agreement which includes pharmaceutical development.

## Is a world IPR system necessary?

Sherwood,[5] among others, has argued in favor of a uniform IPR system for the entire world. Chomsky,[6] on the other hand, has contended that a global agreement under GATT would lead to a greater human cost and a monopoly by the developed nations with developing nations trapped in a situation of no progress. He has written that patents are designed to insure that the technology of the future is in the hands of transitional corporations.

The assurance that others can be prevented from unauthorized copying has become a powerful stimulus to the process of invention, technical advancement and creative expression. If a congruent world intellectual property system were to be successfully instituted, it would provide a powerful stimulus to the global economy. It would lead to increased investment and international cooperation in research and development and hopefully increased transfer of technology between nations. However, several practical problems need to be overcome before such an ideal state of affairs can be achieved. These include:

(a) A basic incompatibility between traditional indigenous practices and IPR regulations; for instance, Park and Ginarte[7] have pointed out that, in some cultures, new ideas and technologies are regarded as "public goods," which should be shared freely by all members of the community. In other words, cultural pride takes precedence over the incentive for private benefits;

(b) A lack of an affordable legal and administrative infrastructure to enforce IPR regulations in several developing countries;

(c) Increased tariffs incorporating the profit system of western countries would be prohibitively expensive for many developing countries;

---

[*]The Government of Costa Rica established a research organization (INBio) to aid biodiversity conservation, which was supported by Merck & Co. with particular regard to pharmaceutical development.

(d) The fear that small regional industries would be unable to compete with large multinational corporations, and so on.

Armstrong[8] has suggested that the best reason for globalizing IPR systems is because research, development and invention are all increasingly becoming international activities. Countries with weak or no intellectual property systems tend to receive less technical knowledge because (a) even if proprietary knowledge is unlawfully copied, the violators will have no access to the associated knowledge to establish a sustained technological base, (b) they lack trained leaders and technicians in the new technologies, and (c) there is a limited market for "pirated" products, especially in countries with strong protective systems. Such situations would not be conducive to foreign investment, which is urgently needed in many developing countries.

The most serious consequence of a weak intellectual property system is the lack of opportunities for participation in the latest global activities in the development of new technologies which often tend to displace entire industries. This process of rapid change is ultimately expected to shape global IPR systems. Such a system should be strong but flexible enough to absorb any surprises which science and technology continue to bring forth. There is great pressure to diminish the time needed for a scientific invention to reach the product application stage. These considerations will continue to shape national and international priorities, as well as impact future IPR systems.

## References

1. M. S. Swaminathan, ed., *Agrobiodiversity and Farmers' Rights* (Konark Publishers Pvt. Ltd., Delhi, 1996).
2. J. Van Wijk and J. Jaffe, *Science Communication* **17**, 338 (1996).
3. M. Khalil, *Intellectual Property Rights and Biodiversity Conservation*, T. Swanson, ed. (Cambridge University Press, Cambridge, 1995), pp. 232–253.
4. T. D. Mays *et al.*, *Valuing Local Knowledge*, S. B. Brush and D. Stabinsky, eds. (Island Press Inc., Washington D.C., 1996), pp. 259–280.

5. R. M. Sherwood, in *Global Dimensions of Intellectual Property Rights in Science and Technology*, M. B. Wallerstein *et al.*, ed. (National Academy Press, Washington D.C., 1993), pp. 68–88.
6. N. Chomsky, *World Orders, Old and New* (Columbia University, New York, 1994), *Science Communication* **17**, 379 (1996).
7. W. G. Park and J. C. Ginarte, *Science Communication* **17**, 379 (1996).
8. J. A. Armstrong, *Global Dimensions of Intellectual Property Rights in Science and Technology*, M. B. Wallerstein *et al.*, eds. (National Academy Press, Washington D.C., 1993), pp. 192–207.

# Biodiversity and
# Intellectual Property Rights (IPRs)

I wonder how many of us realize that the biodiversity of many developing countries is being appropriated by the rich nations in the name of Intellectual Property Rights (IPRs). Biodiversity is the sum total of living natural resources including all of the plant and animal species. This is an urgent problem, yet few individuals other than the experts are aware of the situation. The point becomes immediately clear when one mentions that, in recent years, multinational companies have been attempting to patent basmati rice, neem tree products, and several other natural resources of medicinal and commercial importance. The medicinal components of these plants have been used for several thousand years by indigenous people for medicinal and nutritional purposes. However, under new patent laws (if accepted), the same products will be too expensive for use by the people in developing countries.

There is also another danger. While commercial profits go to the rich countries, the natural resources of developing countries such as India are being slowly eroded. Unless conservation measures are taken immediately, several of these valuable species will become extinct.

In a meeting with the Indian President, K. R. Narayanan in New York in 1998, I discussed this problem. He referred to this matter in his acceptance speech to the Appeal of Conscience Foundation when he was awarded the World Statesman Award. It was an impressive ceremony at the New York Hilton, where President Narayanan was introduced to the distinguished gathering by Dr. Henry Kissinger. Although during the Nixon administration, Kissinger was advocating a "tilt" toward Pakistan, he is now a strong advocate of U.S.–India trade. His firm, the Kissinger Associates, is actively involved in promoting U.S. trade with China and India, the two most populous nations of the world. In his speech, President Narayanan stated, "The prospect of a technological elite emerging in industry, information and in the crucial areas of life is a new possible danger to the independence, initiative, and the identity of the individual human being. It seems that the human race has to guard against the rise of high tech hegemonism...

And yet, I believe that when one compares side by side, the increase of the power of the rulers and the economic and technocratic hegemons, and the increase of public consciousness, the consciousness has become an even greater force in society... This is the greatest safeguard against a techno-financial hegemony."

With the rise of international technocracy (which is clearly controlled by the rich nations), Intellectual Property Rights and international patents have become the rule. How can impoverished nations compete with this situation? One way is to put forward alternative proposals under the WTO (World Trade Oganization) under the auspices of the United Nations. This is being pursued actively by several countries. The other method is to educate the public all over the world to sensitize them to the disparities in the IPR laws, which have been largely promulgated by the developed countries. President Narayanan emphasized public awareness as the most urgent and plausible remedy in democratic countries. In this respect, both the United States and India have much in common, sharing the founding principles of religious freedom, tolerance, equality and justice, and with consideration for human rights and human values.

# *Patenting in Biotechnology*

## Patenting in biotechnology

What is biotechnology? It may simply be defined as the application of biological knowledge for the purpose of solving practical problems in healthcare, agriculture and related fields. Recent years have seen an explosive growth of research in biotechnology and its applications. With the advent of new inventions and discoveries in genetic engineering, numerous patents in biotechnology involved a genetically engineered bacterium, Pseudomonas aeruginosa, which was designed to break down four of the main components of crude oil. The patent was approved in 1980 by the U.S. Supreme Court in a 5–4 decision in the case of Diamond versus Chakrabarty. This marked a new beginning for patenting "man-made" living organisms. It was based on the premise that the patent legislation, which was earlier enacted by the U.S. Congress, did not distinguish between "living" and "non-living" matter. The Chakrabarty patent opened the door for numerous other patent claims in biotechnology in the following years.

## Criteria of patenting

There are three major criteria which need to be fulfilled in order to satisfy patenting regulations: (a) novelty (not previously known), (b) nonobviousness (invented by human ingenuity), and (c) utility (practical application). There is a widespread belief that patenting genes does not meet the criteria required. A gene may not be known previously, but the process of making it known can be regarded as a discovery, not an invention. This leads to the question: "How can you invent something that already exists in nature?" It has been suggested that the skills required to construct full-length genes and define their function and utility are not straightforward, requiring a certain degree of inventive-ness. The second major objection is that the mere process of knowing a sequence of genes does not reveal its utility. I will return to this topic later.

## International treaties

The Paris Union Convention Treaty was signed on March 20, 1883, and entered into force on July 7, 1884. It has been revised at least nine times since then. It is a universal (worldwide) treaty, establishing certain basic rights for the protection of property and covering a wide variety of industrial patents.

## The Budapest Treaty

Patenting micro-organisms and cells is not a recent phenomenon. Louis Pasteur was awarded a patent involving a living matter-yeast, in 1873. More recently, the "Budapest Treaty" on the International Recognition of the Deposit of Micro-organisms for the purposes of patent procedures became effective on August 19, 1980. Its major aim is to provide recognition, for the purpose of their own patents, by the member states of a deposit of the micro-organism strain which is made in another country of the Treaty. Its provisions include a series of International Depository Authorities (IDA) which are depository institutions located in a member country and are recognized by the appropriate national or international organizations (WIPO) by which the institution guarantees compliance with a number of regulations as required by the Treaty. However, the treaty does not specify the details of the micro-organism deposits. These are left to the discretion of the domestic laws of the relevant country.

# Patenting in Agriculture and Medicine

During the years following World War II, many nations began to lower their trade and tariff barriers to make them more open and accessible for international trade under the auspices of the General Agreement on Tariffs and Trade (GATT). To reduce the border barriers and restrictions for trade, multilateral negotiations have been taking place. For instance, the Kennedy Round in the 1960s, the Tokyo Round in the 1970s, and later, the Uruguay Round in the 1980s, which was formally signed in April, 1994.

The Uruguay Round included, for the first time, an active participation by developing countries. One result of the Uruguay Round negotiations was the stipulation that all signatories should adopt trade-related intellectual rights protection (TRIPS). Another development was to achieve agreement that trade in services would be brought under multilateral trade disciplines in the World Trade Organization (WTO). Simultaneously, the Convention on Biological Diversity (CBD) was negotiated before the United Nations Conference on Environment and Development (UNCED), which was held in Rio de Janeiro in 1992. The first meeting of the conference was held in the Bahamas in 1994. It was a moment of great importance to the developing world (formerly known as the "third world countries"). Article I of the convention stated the objectives to include the "conservation of biological diversity,... the fair and equitable sharing of benefits arising out of the utilization of genetic resources and by appropriate transfer of relevant technologies..." Article II defined "biological diversity" as the variability among living organisms from all sources at all levels. To put it simply, biodiversity is an index of the biological wealth of our planet. The Global Biodiversity Convention, which became effective on December 29, 1993, recognized that states have sovereign rights over their own biological resources and are responsible for conserving their biological diversity and for using their biological resources in a sustainable manner.

## Intellectual property rights

An outcome of these considerations is the need to protect the ownership of specific crop varieties and plant genetic resources. The germplasm of many naturally occurring crop varieties is the end product of long periods of evolution. Unless protected and conserved carefully, these genetic resources can easily be lost. Farmers in developing countries (where much of the naturally occurring biodiversity originates) have largely been responsible for preserving and propagating these species over centuries. Yet modern intellectual property laws, which are mainly designed by the developed countries, have failed to recognize their contributions. In addition to the lack of recognition of intellectual property, the monopoly situation created by multinational companies, through the patenting process, poses potential dangers by controlling food shortages and by increasing prices; it may even result in the extinction of rare species. A prime example is the development of the "terminator" gene in cotton by Monsanto and its partners. The gene produces a crop with inviable seeds which will be useless for growing the next crop. Consequently, the farmer will be totally dependent on the company that supplies the seed for growing every crop. It would be a monopoly controlled by Monsanto and its partners. If the gene can be introduced into all major crops worldwide, it could lead to total control of the world's food production and supply by Monsanto and its partners. There have been numerous protests by Indian farmers against this possibility. A related problem is the lack of recognition of indigenous knowledge and contribution which has conserved and protected the medicinal plant species of Asia, Africa, and Central and South America. Although some western pharmaceutical and drug companies have benefitted from that knowledge, the tribal and aboriginal communities who preserved these resources for centuries have not received a share of the profits.

## Human biodiversity

The human genome project has been producing revolutionary new data on the nature and location of a great number of genes which cause many diseases, including various types of cancers. This explosion of new data has also resulted in several new controversies and ethical dilemmas. For instance, is it ethical to patent the discovery of new genes (pieces of DNA), which are only the naturally occurring parts of our body cells? Some scientists have suggested that a gene is patentable only if its study has led to the development of a cure for a disease. Others have felt that the discovery of a gene for a specific disease (such as breast cancer) by itself merits patenting. Several feminists have challenged the patenting of the breast cancer gene, which they have regarded as a natural part of the body of the human female. This debate and other similar challenges are continuing. There are also other ethical problems such as the privacy and confidentiality of private genetic information about individuals. Where should we draw the line in making such information public? This information could adversely affect an individual's employment opportunities and health insurance status. A similar problem is related to the confidentiality of the genetic profiles (DNA fingerprinting) of individuals that are being collected by the FBI and other law enforcement agencies. Several civil rights organizations such as the ACLU have expressed concern about their disclosure. This brief discussion has given some idea of the ramifications of the ethical dilemmas that are resulting from the application of intellectual property laws and the patenting process, both at the individual level and at the national and international levels.

# Should We Take an Interest in Intellectual Property Rights?

There is an old song from a Raj Kapoor movie which goes something like this: "My shoes are Japanese, my pants are English... but my heart is Hindustani." No matter which part of world we came from, most of us have sentimental feelings about our motherland and express our feelings in different ways. Each of us, in our different ways, maintains close ties with our friends and relatives back home.

As a scientist, I have been trying to promote the transfer of science and technology to India for several years. In 1994, I initiated the JBS Haldane International Symposia which take place at a center in India each year. The third Haldane Symposium was held in New Delhi on January 2, 1997. That year's topic was "Intellectual Property Rights and International Cooperation in Science and Technology." The co-chairmen of the symposium were myself and Dr. R. A. Mashelkar, Director-General of the Council of Scientific & Industrial Research, and Secretary to the Governor of India, in New Delhi.

The world is now shrinking and what we do here in America affects other countries; especially as Indo-Americans, our actions have a strong impact on India. Under the new GATT regulations, particularly since the Biodiversity Convention in Rio de Janeiro in 1992, developing nations have become much more aware of the consequences of not having a formal policy toward intellectual property rights. One example illustrates the situation. For the past several years, the following scenario has been taking place.

Rich countries have been sending exploratory teams to the forests of poorer nations in Asia, Africa and South America for the purpose of screening and exploiting medicinal plants and other plants which have various commercial applications.

Several problems become evident in this context. Firstly, the traditional healers and indigenous people who have preserved ancient knowledge do not receive a share of these benefits when big pharmaceutical companies extract and patent drugs from the natural resources of developing countries. Secondly, over-exploitation and neglect have led to the extinction of many plant species. The IUCN (International Union for the Conservation of Nature) and the WWF (World Wildlife Foundation) have estimated that 60,000 higher plant species could become extinct by the middle of the next century unless adequate conservation measures are implemented immediately. Thirdly, it is estimated that drugs containing one or more plant-derived active ingredients represented about 25% of all prescriptions dispensed from community pharmacies in the United States. In Eastern Europe, the proportion of prescriptions containing plant drugs is more than 60% because traditional medicine and modern medicine are amalgamated. The traditional Indian system uses more than 200 species of plants in their pharmacopoeias. Lastly, however, short-term exploitation by multinational companies which have no conservation plans or reforestation programs will certainly lead to the rapid depletion of valuable natural resources from developing nations.

One remedy to correct this trend is to institute a systematic policy to protect resources by establishing intellectual property laws which will require just compensation for any export of biological materials or other valuable resources. For several centuries, foreign powers have practically stolen the natural wealth of developing countries, but paid no compensation until now. That was the situation under colonialism in the past. Now, multinational companies (which are mostly American, but some are also European) are repeating the same exploitation, taking advantage of local communities who possess little knowledge of their natural wealth. For a few dollars, such multinational companies have been able to make big profits by exploring natural plant resources (e.g. the *neem* tree) from poor countries.

The new GATT treaty, which incorporates Trade Related Intellectual Property Rights (TRIPS), is controversial. Some see it as a path to global liberalism, which will restore world order to address political, economic, social, environmental, legal and scientific concerns. Others see it as a new form of exploitation. Prominent social scientist Noam Chomsky of MIT in Cambridge, Massachusetts wrote: "It is a scam designed to rob the poor and enrich the rich, like most social policy... people suffer."

# Drug Development and Intellectual Property

Shaman Pharmaceuticals in San Francisco, California, is an unusual company. For the past several years, it has been signing agreements and collaborating in medicinal plant research with tribal and aboriginal peoples all over the world. What is unusual about their work is that they have made a sincere effort to share the benefits of drug development with developing countries in a fair and equitable manner; this has not been the case with other companies. In addition to sharing financial benefits, Shaman's program also includes a recognition of the intellectual property of the tribal and forest-dwelling peoples of Asia, South America and Africa. It is well known that some multinational companies from Europe and North America have been illegally patenting the intellectual property of certain developing countries; for instance, W. R. Grace's patenting of the medicinal properties of the *neem* tree. Products from this tree have been used therapeutically in India for several centuries. The claim of intellectual property in this instance by W. R. Grace & Co. is not valid.

## Reciprocity

Shaman Pharmaceuticals has demonstrated good faith in its dealings with the tribal communities of the developing world. There is usually a lag of five to ten years to develop a marketable drug (if successful) from plant products. Shaman established trust with the indigenous communities immediately by offering certain benefits at the beginning of their association. The nature of these benefits depended on the needs of the tribal population involved. In December, 1991, Shaman began discussions with the Quichua Community in the Ecuadorian Amazon. Upon request, this group listed a number of items of immediate need. These included: (a) building a new airstrip in the jungle for emergency medical evacuations, (b) quarterly visits by a culturally sensitive physician and dentist, (c) providing funds for training young people in the community, (d) providing a supply of over-the-counter medicines such as aspirin, and (e) providing employment to several people in the community. These benefits were provided long before any drug was developed. It was not even clear at that stage if there was going to be a successful outcome of their research program to produce a marketable drug. In addition, Shaman signed an agreement to share all research data with the community and any potential financial benefits from future drug sales. This is a long-term continuing relationship, which includes both intellectual and physical property. Another case involves collaborative drug research to discover and develop lead anti-viral products, Provir and Virend, and included several countries in Latin America. A major concern of this group was a proper acknowledgement of the source of the medicinal plants.

Shaman signed similar agreements with thirty communities in developing countries. Besides sharing the intellectual property as well as the financial benefits, many communities have expressed a keen interest in the conservation and protection of medicinal plant species. Training programs and supply of scientific equipment and medical stockpiles are additional benefits. The world's greatest biodiversity is found in developing countries. Much of the drug development in North America and Europe would not be possible without the biodiversity resources of the poor countries in Asia, Africa and Central and South America. Among the developing countries, Indonesia, India, Thailand, Sri Lanka, Vietnam, China, Nigeria, Botswana, Colombia, Mexico, Ecuador, Peru and Brazil have the greatest diversity of plant species in the world. But many developing countries are only just learning the value of their own biodiversity and the urgent need to protect that property.

International organizations such as the "International Union for the Conservation of Nature" (IUCN) and UNESCO have been promoting this awareness through public education and research funding. Billions of people in developing countries depend on indigenous medicine, such as the Indian Ayurvedic medicine, for their primary care. This tradition has sustained human populations for several millennia. There is an urgent need to protect the intellectual knowledge and plant resources which have provided the foundation for these ancient traditions.

# GLOBAL ISSUES
# IN BIOTECHNOLOGY

# Global Economy and Genomics

Genomics is the art and science of understanding the genetic structure of living organisms, each cell containing a set of DNA instructions. It encompasses a great deal of information such as the identification and location of specific genes and their sequential arrangement on chromosomes in the cells, the entire package being called "the genome." In recent years, genomics has been used to apply valuable genetic information in biotechnology, agriculture and pharmaceutical industries. Genomics has become one of the biggest sectors in the global economy. Some of the world's largest multinational companies are restructuring themselves to replace their traditional activities in chemicals, fertilizers and energy industries with genomics-based enterprises. The boundaries between agriculture, pharmaceuticals, cosmetics, health, environmental and nutrition-related industries are disappearing.

## Company mergers

Large companies are trying to survive by forming mergers and alliances. In so doing, they are increasing their research budgets. The merger between the two Swiss pharmaceutical giants, Sandoz and Ciba-Geigy, created a US$100 billion conglomerate known as Novartis. In December 2000, Glaxo Wellcome and SmithKline Beecham joined forces to become GlaxoSmithKline with a market capital budget exceeding the combined annual budgets of over a hundred of the world's nations. Several companies are also creating complex multiple alliances and partnership networks. According to Juan Enriquez of the David Rockefeller Center in Cambridge, Massachusetts, alliances between small drug companies and large pharmaceuticals grew almost six-fold between 1993 and 1996. These alliances provide small companies with access to large laboratories, databases, and technical guidance while maintaining their entrepreneurial independence.

Novartis recently announced that it will spend US$250 million on functional genomics, combining DNA chips, bioinformatics and cross-species research. The new Genomics Institute will promote molecular epidemiology, structural biology, bioinformatics, combinatorial chemistry, genetic screening, transgenics and so on. Merck has also been investing heavily in research. While seventy biotech firms had 284 drugs in development (with a market capital of US$50 billion), Merck had 26 drugs in development and a market value of US$80 billion. Glaxo-Wellcome plans to create a Genomics Institute with a budget of US$47 million. SmithKline invested US$125 million in the genomics industry, gaining exclusive rights to the database from the Institute for Human Genome Sciences.

Several chemical giants such as Monsanto, Dow Chemicals and Dupont are reinventing themselves as life science companies. For over ten years, Monsanto has been getting rid of its various chemical and energy divisions and has, since 1997, invested US$6.6 billion in biotechnology and genomics. Its future plans include an even greater investment in biotechnology and agribusiness. Investment in pharmaceuticals is greatly expanded through its subsidiary, G. D. Searle & Co.

## The terminator gene

In May, 1998, Monsanto bought DeKalb Genetics and Delta & Pineland for US$4.2 billion, creating a joint venture with Cargill, one of the world's largest seed supplying companies. Its major objective is the commercial exploitation of genetically engineered foods and crops. One consequence of their research program is the so-called "terminator" gene, which has caused much consternation throughout the developing world. Several farmers in India are supposed to have committed suicide because of the imminent contamination of crops by the terminator gene.

It is so named because the gene causes the crop to produce inviable seed which fail to germinate and hence there is no crop next season. In other words, farmers have to buy fresh seed from the seed companies each year. In developed countries, the terminator gene will force farmers to plant improved seeds in each generation. Farmers from the developed countries will be able to afford them, but it will be an unbearable burden for those in the developing countries.

Monsanto's success was imitated by DuPont and Dow Chemicals. DuPont announced in 1998 that it would be divesting the energy company, Conoco, and investing in a life sciences company. Its stock immediately went up by twelve percent. Dow Chemicals paid US$900 million for Eli Lilly's 40% share in a joint venture to modify crops and foods. Hoechst too got rid of its basic chemicals divisions, investing in pharmaceuticals and biotechnology, and joined Schering in an agribusiness venture. In addition, it bought Plant Genetic Systems for US$600 million.

Future mergers will result in new combinations of medicine, chemicals and food. Medical prescriptions will appear in the form of genetically altered foods. Some vaccines will be delivered through modified potatoes or bananas. Companies that produce pharmaceuticals may merge with cosmetics manufacturers to produce "cosmeceuticals." Proctor and Gamble and Archer Daniels Midland will form alliances with genomics firms which produce cosmetics and foods, and will deliver genetically modified products for treating various physical and mental disorders. One of their first goals will be to reverse the aging process. Other targets will be heart disease, cancer and common allergies. Energy companies may learn to produce energy from genetically modified plant sources. The future holds many promises and surprises, which will come about through the successful mergers of companies and multiple technologies.

# Risks Resulting from the Application of Agricultural Biotechnology: What is the True Situation?

A great number of real, imagined or potential risks to the population have been attributed to the application of agricultural biotechnology. Future research and experience will enlighten us as to whether the proponents of such dire predictions deserve to be lauded as heroes or mere false prophets.

To understand this point of view, I will start by listing some of the hazards mentioned by various writers, such as Mae-Wan Ho, regarding agricultural biotechnology.

It has been said that agricultural biotechnology does not feed the world, is unsustainable, and poses unique hazards to health and biodiversity. The main point of this argument are the following:

(a) the increased drain on genetic resources from the South to the North;

(b) restrictions resulting from the implementation of intellectual property rights and the large scale control of food production by corporations, which are increasingly marginalising family farmers;

(c) the gradual loss of traditional technologies;

(d) the inherent genetic instability of transgenic crops;

(e) toxic effects caused by the interaction of transgenic products with the host genome;

(f) increased use of pesticides with pesticide resistant transgenic crops, leading to ill-health and water contamination;

(g) the spread of antibiotic-resistance marker genes to pathogens by horizontal transfer and recombination;

(h) the spread of virulence among pathogens between species;

(i) the potential for creating new pathogenic bacteria and viruses;

(j) in transgenic foods, the potential for transgenic DNA to infect cells after ingestion, causing the regeneration of disease viruses including cancer;

(k) the spread of transgenes to related weed species, creating herbicide superweeds;

(l) the indiscriminate use of herbicides with herbicide-resistant transgenic plants leading to the elimination of indigenous species;

(m) the increased exploitation of natural biopesticides in transgenic plants, creating a corresponding range of resistant insects, depriving the ecosystem of its natural pest controls; and

(n) the potential for transgenic DNA, unlike other forms of contamination, to unleash cross-species infections that would be difficult or impossible to control.

I will consider two specific examples. There is recent work which provides evidence that pollen from *Bt* transgenic corn kills larvae of the monarch butterfly, *Danaus plexippus* (a difference of 20% mortality as compared to 0% when fed near nontransgenic corn). Even a low density of pollen grains caused significant mortality. However, there was no effect on surviving butterflies' development. Clearly, this is an example of an unexpected adverse consequence of genetic engineering, which involves a major crop.

Some other claims remain unproven and speculative so far. There is no evidence showing that transgenic DNA has caused cancer in any human being. Other situations are being constantly monitored and screened for the possibility of toxic effects.

However, the current situation regarding world biotechnology is far from satisfactory. Implementation of intellectual property rights are enriching rich countries at the expense of developing countries. Biodiversity is being eroded all over the world because of increased human populations and their need for survival. Developing countries contain much of world's biodiversity. Much of their traditional medical knowledge and medicinal plant species are disappearing. This has resulted in increased dependency on more expensive western medicine.

## The Technology Protection System (TPS) (terminator)

Three main problems can be listed as stemming from transgene technology: (a) the risk of transgenes spreading to wild species; (b) the need to protect the technology from unauthorized use; and (c) the need to meet the cost of technology development. The system, known as the Technology Protection System (TPS) or popularly known as the so-called "terminator," is designed to meet these requirements. Although the seeds produced by an activated TPS plant would be fully developed and can be used for cattle feed or food grains, they are not capable of germination. In addition, the pollen produced by the TPS plant, if it fertilizes an ovule on a neighboring plant (whether it is a crop or a weed species), will produce seed that is not viable. In other words, the TPS genetic mechanism does not contaminate the neighboring plants. So it is useful for testing transgenic crops or other experimental crops in limited localities in environmentally sensitive areas.

The impact of having TPS in a genome is not irreversible. Its mere presence in the genome does not affect the plant. It is only activated when the seed is treated with low levels of tetracycline. (This treatment activates the site-specific recombinase gene during germination, leading to the removal of a DNA blocking sequence located between the late embryogenesis promoter and a gene that disrupts the germination process.) It is only at the late stage of seed development that the seed cells produce the protein synthesis inhibitor protein.

Mechanisms such as the TPS will be required in the future as research is now expanding to use various crops and plant species for the production of biopharmaceuticals. These offer better alternatives than the more traditional expression systems such as mammalian cells. Most of the major crops are now amenable to transformation because increasing knowledge about promoters and targeting sequences is facilitating transgene expression to be specifically targeted to storage tissues for easy storage and isolation. Compared to mammalian cells, plants produce high yields of protein at a fraction of the cost. Furthermore, they are safer because the possibility of contamination by human pathogens in animal cells does not exist in plant tissues. Several biotechnology companies are now actively developing edible vaccines which can be produced in tubers, fruits and seeds. This could bring down the cost of pharmaceutical products. One example is a recent report of oral immunization by the production of an edible hepatitis B surface antigen in potato tubers.

## The future of biotechnology

The situation of biotechnology in relation to agriculture can be summarised as follows:

(a) producing more food on the same area of land, or even less area in some localities, (preserving wilderness, rain forest or other marginal lands);

(b) realizing postharvest loss of food by protecting it from various pests and other natural calamities;

(c) improving the nutritional content of food and even producing foods containing biopharmaceuticals by means of genetic engineering, targeting such diseases as diabetes and heart disease;

(d) using genetic engineering methods to reduce dependency on fuel, fertilizers, pesticides or other traditional resources, leading to an improvement in environmental quality and reduced cost of production;

(e) improving the quality and yield of certain species which are particularly important for human society, such as medicinal plants; and

(f) exploiting plant species for the production of specialty materials, such as polyhydroxybutyrate, a biodegradable polymer.

# Persistence of GMOs in the Environment

The widespread use of Genetically Modified Organisms (GMOs) in recent years has not resulted in any obvious accidents or pathological consequences. However, certain concerns continue to persist in our society today. For instance, horizontal gene transfer is recognized as the main avenue of exchange of genetic material and perhaps also of the spread of antibiotic resistance genes. The question then arises whether the increase in the frequency of outbreaks of new and emerging infectious diseases is, in any way, related to the extensive use of antibiotic resistance genes in genetic engineering contributing to the increase in frequency of antibiotic resistance in bacterial pathogens. What would be the consequence of a spread of recombinant genes in a population? We do not yet have any specific data on the impact of recombinant DNA on a population with respect to a specific disease under various environments nor its long term consequences over several generations.

In recent years, many individuals have repeatedly sounded the alarm about the potential adverse impact of GMOs on our health and environment. Such eminent agricultural scientists as M. S. Swaminathan of India have been advocating, for the past several years, a cautious approach to the use of modern biotechnology from the viewpoint of ecology and health: "...Past experience with new agricultural technologies in India reveals that it is important to subject *new tools and techniques* [italics added for emphasis] to a proactive impact analysis from the points of view of ecology, economics, equity and employment, if they are to lead to sustainable improvements in the quality of human life." (See Foreword to *Biotechnology in Agriculture: A Dialogue*, ed. M. S. Swaminathan, Macmillan India Limited, Madras, 1991).

Older GMOs have been around for over 25 years. Younger GMOs, such as those used in maize and soya in the USA, have been in crops covering millions of acres each year. Yet, there have been no reports of any major catastrophes so far. However, there are several related issues that are of major concern.

Bacteria are known to survive and conjugate in diverse environments, in the digestive tracts of vertebrates and invertebrates, in wastewater and sludge, in aquatic ecosystems, and in soil. Bacteria taken from their environment and released after transformation can re-establish themselves in their original ecosystem. This has been demonstrated in marine habitats.

Spores of genetically modified strains of *Bacillus subtilis* have been shown to persist for prolonged periods in sterile soil, often withstanding stress factors such as "hunger" by reducing their cell size. They revert to their normal state when favorable conditions return. GMOs have been shown not only to survive, but also spread in soil, increasing the chances of gene transfer. However, the standard techniques employed for counting the numbers under different conditions at various soil layers are not as efficient as they should be. They are, at best, gross underestimates.

## Plants

The persistence of GMOs on plants seems to vary according to the species of plant studied. For instance, populations of *Pseudomonas* maintained the same density on leaves of capsicum and egg plants for prolonged periods, but decreased on leaves of strawberry and tomato. Genetically modified strains of *Pseudomonas fluorescens*, derived from parental strains from sugar beets, persisted on sugar beet leaves for almost two years. They were also found on sugar beets planted the following season, spreading to other plants and the soil.

Studies of the survival and persistence of bacteria in their normal as well as dormant states has received much attention in recent years.

Viable non-culturable cells do not lose their plasmids,[*] but may even enrich their intracellular plasmid content. These dormant cells becomes active again under favorable environmental conditions, involving fluctuations of pH, temperature and nutritional supply. Adaptation of GMOs to environmental fluctuations is a slow process, but their numbers can increase rapidly afterwards under the pressure of natural selection. Quite often, GMOs may only require a brief period of survival to be able to be carried to other niches or to infect neighboring cells. Only living cells are capable of conjugation and transduction; however, transformation may take place after the death of donor cells. The probability of gene transfer depends on the quantity and stability of isolated DNA. This is of common occurrence, as seen from the abundance of extracellular DNA in the surrounding environment. However, plasmid incompatibility may hinder genetic transfer.

## Genetic transmission

The spread of antibiotic resistance markers in bacterial communities shows that gene transfer occurs frequently *in vitro* as well *in vivo*. The following are some of the mechanisms of transfer.

### (a) Conjugation

In a bacterium such as *E. coli*, conjugation is the effective means of genetic transmission under natural conditions. Its occurrence appears to be ubiquitous under almost any kind of environment and it happens more frequently under certain conditions. Some of these circumstances are environments with high densities of potential bacterial donor and receptor cells, as in digestive tracts and wastewater treatment plants, to name two entirely diverse environments which promote conjugation. Other factors of influence include soil conditions, temperature, soil

---

[*]Plasmids are extrachromosomal genetic units that replicate in step with the bacterial chromosome. About 1% to 10% of the genome of many bacterial species are plasmids. They can be easily transferred or lost by conjugation, transformation or transduction. The traits specified by plasmids are antibiotic resistance, toxic heavy metal resistance, symbiotic and virulence determinants, resistance to radiation, etc.

humidity and supplements of plant material. In addition, heavy metals, herbicides and antibiotics could either promote or inhibit conjugation in natural conditions.

## (b) Transduction

Transduction contributes to the transmission of plasmids in living cells, mediated by a bacteriophage or phage, when conjugation cannot occur. It is more likely to occur in environments of high cell density and abundant nutrition. Phage infections begin with adsorption of the virus particles to specific receptor sites on the surface of the cell, which increases the chances of survival and transduction.

## (c) Transformation

Transformation does not depend on living cells, but requires the active uptake of isolated DNA by recipient cells which are in an appropriate physiological state of competence as determined by growth stage, impact of stress factors and the species in question. This leads to the integration of the new sequences into the genome of the host cell (recipient) by means of recombination. Homologous sequences have better chances of transformation, which occurs frequently in natural environments. Microorganisms in diverse environments can be transformed by isolated DNA. Feeding with GMOs has been shown to lead to the transformation of endogenous bacteria in the gastrointestinal tract of *Folsomia candida* after feeding with GMOs. Transformation is known to be influenced by the abundance of nutrients, certain minerals, ionic strength of water and temperature.

Biological containment, employing safety vectors but without transfer genes, is achieved by designing special plasmids. On the other hand, shuttle vectors and specially constructed plasmids are capable of transfer across species, orders and higher classes, to ensure their persistence and integration.

# GMOs: *Safety Measures*

The concept of biological containment is the foundation for enforcing regulations concerning the use of Genetically Modified Organisms (GMOs). It has been claimed that the DNA ingested with food is not completely degraded in the gastrointestinal tract, but may enter white blood cells, spleen and liver cells. Ingested DNA has been shown to be transferred to the cells of foetuses in newborn mice (Doerfler and Schubbert, 1997).

Special constructs of nucleic acids are being designed to promote effective replication in different kinds of cells and to reinforce integration and stability by means of recombination. However, these laboratory models do not necessarily reflect what might actually happen in natural habitats in human organs or in the environment. Further research is required to understand the effect of various stresses and extremes of environments upon the stability and dispersal of recombinant genes that are embodied in GMOs.

## How long does DNA last?

Several chemical reactions, such as hydrolysis, alkylation, UV radiation as well as oxidation, have been cited as factors that lead to a progressive fragmentation of the double helix into small fragments. Both temperature and pH of the medium have been shown to affect this process. The half-life of DNA varies significantly in different environments.

**Table**

| Location | Half-life of DNA |
| --- | --- |
| Wastewater | 0.017–0.23 |
| Freshwater | 4.2–5.5 |
| Sea Water | 3.4–83 |
| Marine sediment | 140–235 |
| Soil | 9.1–28.2 |

(After Rollo, 1998, in *Advances in Molecular Ecology*,
ed. Gary R. Carvalho).

Different components of DNA and proteins are absorbed by the
particulate constituents of soils, sediments and clay minerals and others
such as the organic compounds can combine with DNA, forming
complex compounds. DNA half-life may be very long in sediments,
particularly when the DNA is protected inside dead cells.

The answer to the question, "How fast does DNA degrade?" is
derived from studies of archaeological and palaentological evidence.
Under certain situations, the original DNA had been identified in
biologically derived materials which date back 50,000 years.

As a general rule, a DNA molecule appears to be very short-lived in
a warm environment that is rich in water, oxygen, and microorganisms.
On the other hand, the same molecule would survive for years, centuries
or even millenia in a cold, dry, anoxic and sterile environment such as
that found in the permafrost region of Siberia. Rollo *et al.* (1994)
attempted PCR amplification of mitochondrial and nuclear sequences
from 1000-year old kernels of corn from a Peruvian site. The amplification efficiency was found to vary greatly with the length of the target
sequence. Over the past several years, various reports of DNA isolation
have been published. They are from paleontological samples that are
many millions of years old, such as dinosaur bones, plant leaves and
insect parts. However, their authenticity has been questioned by others.
An investigation of amber-entombed insects also produced no reliable
results because of the failure to amplify authentic insect DNA. Rollo
(1998) concluded that the upper temporal threshold for DNA survival
lies between 50,000 and 100,000 years.

## Plasmids and horizontal transfer

Many genes of great importance in medicine, agriculture, environment and industry are located in bacteria plasmids, yet we know little about their characteristics, host range and habitat, or their genetic diversity. With respect to the role of plasmids in the spread of GMOs, the following points are of interest:

(a)  new methods and tools for characterizing plasmids;
(b)  the size and composition of the horizontal gene pool in the bacterial genome;
(c)  the normal host range of the major groups of plasmids in a range of bacterial communities in diverse environments;
(d)  extending our knowledge of plasmids to a broader section of the prokaryotic kingdom; and
(e)  genes carried in the horizontal gene pool.

By demonstrating the important role that plasmids play in bacterial population genetics, it should be possible to devise a method for utilizing certain types of GMOs without undue risk to the environment and health. Predictive models need to be developed on the basis of the molecular and environmental factors that initiate transfer.

It is difficult to predict the exact circumstances that might determine the fate of a transgenic GMO in a population. It is well known from the pioneering mathematical analyses of J. B. S. Haldane and others that it only takes a slight selection pressure to cause the spread of a rare recombinant gene in a population. While one must explore all possibilities, it is clear that the theory of evolution and selection is totally inadequate when considering the fate of GMOs. Under these circumstances, it is perhaps correct to conclude that there is no absolutely safe GMO that can be relied upon under varying biological and environmental conditions.

## National Academy of Sciences

In 2000, a committee of the U.S. National Academy of Sciences issued a comprehensive report on genetically modified pest-protected plants, assessing the risks to health and the environment as well as the regulations that needed to be modified to protect the public. Among the committees' evaluations, health impacts of GMOs were considered in terms of allergenicity, toxicity and pleiotropic effects. The committee recommended that priority should be given to the development of improved methods for identifying potential allergens in pest-protected plants, specifically, the development of tests with human immune system endpoints and of more reliable animal models. The committee emphasized the need to collect long-term data on animals feeding on transgenic pest-protected plants because of its potential value in assessing the impact on human health.

The committee found no evidence that foods on the market are unsafe to eat as a result of genetic modification. Among other aspects, the committee suggested further studies to determine the impact of specific pest-protected crops on nontarget organisms, assess gene flow from transgenic crops and its potential consequences, and monitor the ecological impacts of pest-protected crops on a long-term basis.

The committee made further suggestions for making the regulatory oversight by the USDA, EPA and FDA more efficient and flexible, and underlined the urgent need to complete the regulatory framework in the near future.

The health impacts of GM technology fall into three broad categories: allergenicity, toxicity and pleiotropic effects. Novel proteins can be analyzed for their digestibility and the degree of protein expression. Homology to known allergens can be evaluated. However, the committee recommended that direct tests should be developed for identifying potential allergens in pest-protected plants in relation to the human immune system. Data are scarce on toxicity testing and pleiotropic effects in relation to introduced gene products. Hence, a primary requirement is to assess data on the baseline concentrations of plant compounds of potential dietary or toxicological concern and how they vary under different specific genetic and environmental conditions. These studies should be undertaken to determine the impacts of specific pest-protected crops on nontarget organisms and the long-term consequences of the impact on animals and their relevance to human health. Furthermore, it is necessary to assess gene flow by listing weedy relatives of major crop plants and identify key factors that regulate weed populations and potential gene flow from crops to their weedy relatives.

# E.U.–U.S. Biotechnology Recommendations

It is of interest that the recommendations of the E.U.–U.S. Biotechnology Forum include a recognition of the contribution of indigenous communities all over the world to agricultural and medical knowledge and the need to compensate them by sharing the royalties in a fair and equitable manner (see Recommendation #21).

Romano Prodi President of the European Commission and President Clinton of the United States agreed, in May 2000, to set up an E.U.–U.S. Biotechnology Consultative Forum to evaluate the issues of major concern in biotechnology in the U.S. and the European Union with reference to the use of modern biotechnology in food and agriculture. The Forum was asked to submit the report to the E.U.–U.S. Summit meeting in December 2000.

The Forum assessed the risks and benefits involved in relation to such factors as health, safety, economic development, food security and environment. Other related issues include public perception, biosafety, risk analysis, intellectual property rights, patenting, ethics and the regulatory process. The biotechnology debate is further complicated by the globalisation process and multinational companies. Its impact on the environment, conservation and sustainable development is another aspect of this evaluation.

The goals and actions of the Forum are stated as follows:

"The Consultative Forum has decided to concentrate its recommendations on the use of biotechnology in food and agriculture...

The scientific world has a responsibility for the public good. The role of science is to serve humankind. Scientists have the obligation to evaluate possible long-term consequences of new technologies and to inform policy makers honestly...

We must not forget that, in the end, it is the public that has to decide whether or not to accept a new technology...

The Consultative Forum has looked carefully at the role of the citizen in relation to both governance and regulatory processes.

The Consultative Forum endorses public responsibility for global governance of biotechnology as one contribution to sustainable agriculture."

## Recommendations

The following recommendations were made by the Consultative Forum in December 2000.

(1) To ensure that genetically modified food and animal feed are safe, all products should be subjected to a mandatory pre-market examination by the appropriate regulatory authorities and approved for sale only after they are found to meet the standard of presenting a reasonable certainty of no harm.

(2) The individuals charged with risk assessment should be well qualified to make decisions in the area under review, be individuals of the highest integrity, and meet stringent requirements for public disclosure of actual and potential conflicts of interest.

(3) More public funds should be invested in basic research that addresses safety concerns.

(4) Consideration should be given to changes in public policy regarding public funding for basic research that would ensure the existence of a vigorous and independent public scientific research enterprise.

(5) The concept of substantial equivalence should only be used to structure a safety assessment. The fact that a biotechnology food is held to be substantially equivalent to a conventional food should not be taken automatically to mean that it needs less testing or less regulatory oversight than "non-substantially" equivalent biotechnology foods. The concept of substantial equivalence should

be improved by the development and application of new techniques, which can help to identify unintended and potentially harmful changes.

(6) Risk/benefit considerations should not be introduced until the basic threshold of reasonable certainty of no harm to human health has been reached.

(7) We recommend that once the basic threshold of human safety has been met it is also appropriate to consider, on a case-by-case basis, the potential risks and benefits of each new product given the health and nutritional status of the people and the ecological and the agricultural systems in a particular region of use.

(8) Governments should undertake to develop and implement processes and mechanisms that will make it possible to trace all foods, derived from GMOs, containing novel ingredients or claiming novel benefits. Before such new products are approved for marketing or when there are sufficient environmental questions, a detailed plan for mandatory monitoring should be established on a case-by-case basis.

(9) There is a need for instruments to enforce effectively the obligation to monitor. For this purpose, the limitation of the duration of marketing approvals may be an appropriate instrument. For these marketing approvals, continued approval would be based upon the results of the monitoring.

(10) A periodic review of the field should be undertaken every 18–24 months by specialists and stakeholders who are responsible for the regulatory process. Mechanisms should also be developed for a way of debating future applications and the issues that they might raise for interested parties at the earliest opportunity in the process. This will help frame the questions that should be addressed by the risk assessors and risk managers.

(11) The E.U. and the U.S. should, as a priority, help to elaborate international rules and procedures in the field of liability and redress.

(12) When substantive uncertainties prevent accurate risk assessment, governments should act protectively on the side of safety.

(13)  All regulatory processes governing the approval of products of agricultural biotechnology should be open, transparent and inclusive.

(14)  The regulatory procedure, including risk assessment and risk management, should include, apart from those usually included (e.g. toxicologists, nutritionists, molecular biologists and plant breeders), a broad range of specialists and stakeholders (e.g. social scientists, ethicists and representatives of civil society).

(15)  Consumers should have the right of informed choice regarding the selection of what they want to consume. Therefore, at the very least, the E.U. and U.S. should establish content-based mandatory labelling requirements for finished products containing novel genetic material.

(16)  The U.S. and the E.U. should commit themselves to stimulating the development of global sustainable agriculture that will provide both adequate amounts and variety of nutrients in a manner that is accessible to all, equitably distributed and culturally acceptable.

(17)  The U.S. and the E.U. should increase public funding in the area of sustainable agriculture and nutrition research in the public interest.

(18)  The E.U. and the U.S. should set up an independently administered fund for the training of developing country nationals in sustainable agriculture, biosafety controls, molecular biology, nutrition and other related fields needed to implement sustainable food production systems, including the effective use of modern agricultural technology.

(19)  The E.U. and the U.S. should pursue the implementation of the biosafety principles outlined in the Cartagena Protocol on Biosafety.

(20)  The E.U. and the U.S. should promote and participate in a global dialogue on an intellectual property rights regime (or some alternate method) that would both provide a fair return on research investment and support sustainable agriculture for the developing world. The aim should be to ensure fair and equitable access for developing countries to new biotechnologies and products. More specifically, developing countries should not be forced to grant

intellectual property rights which could prevent farmers from freely replanting seeds or public breeders from freely using varieties as initial sources of variation.

(21) The E.U. and the U.S. should call for respect of the traditional or indigenous agricultural and medical knowledge in any country of the world and for the fair distribution of the royalties and other rewards from inventions based on this knowledge.

(22) The E.U. and the U.S. should examine the development of incentive mechanisms to encourage private companies to engage in research of particular importance for developing countries, and to make available research results including proprietary technologies to those countries.

(23) We urge the E.U. and the U.S. to promote a transatlantic process for engaging a broad range of stakeholders to examine ongoing issues of biotechnology.

# GMOs *and* Politics

One of the hottest controversies that emanated from the application of genetic engineering is the potential risk involved in the utilization of genetically modified organisms and cells in agriculture and medicine. The debate has become needlessly complicated, acquiring a false sense of urgency, mainly because of its politicization by various groups. Issues of purely scientific interest have been overshadowed by political and social concerns. Instead of an objective and dispassionate evaluation, an atmosphere of mistrust and commercial exploitation has prevailed. It is further evident that examples of practical evidence which can clearly demonstrate the adverse impact of genetically modified organisms or cells on the environment are few and far between. Only further research will enable us to decide whether the application of genetic engineering has resulted in deleterious consequences to humanity. However, the evidence so far available indicates no major disasters resulting from GMO technology.

The anti-GMO faction suffered a major setback recently when the founder and former President of Greenpeace, Patrick Moore, changed his stance dramatically and issued a statement supporting genetic engineering research in agriculture. He stated that "we don't really know of any negative aspects for GMOs, but we do know of many positive ones, both socially and environmentally."

## The Green Party

In recent years, the growing strength of the Green Party has resulted in a stronger opposition to the use of GMOs in Europe than in North America. The Green Party emerged in Britain in 1973 and by 1989 the U.K. Greens had captured 15% of the votes in European elections, but secured not a single seat in the European Parliament. However, in the following years Green Party candidates in France, Belgium, Germany and Georgia captured several seats in their national parliaments and local councils. Gradually, their strength grew all over Europe, leading to the formation of the European Federation of Green Parties in 1993. The "Guiding Principles" of this Federation include eco-development, ecological sustainability, equity, justice, non-violence, racial and gender equality, and nuclear disarmament, among others. The party platform also includes an opposition to DNA technology, genetic engineering, GMO research and its field applications.

The Federation's primary goal is to promote an environment-friendly economic development policy that will benefit Europe. Their strongest representation is found in Germany, with 49 members of Parliament — the third strongest political force in the country. The total number of Green parliamentarians in the European Parliament number over 300 in 2001. They have also made rapid strides in Taiwan, Australia and the United States in recent years.

## The Green Party's manifesto on GMOs

At a Party convention held in Stockholm in November 2000, the European Greens called upon the European Parliament to remove all GM foods from EU markets, ban their further introduction and to resolve that no GM food will be approved in the future unless it has been shown to be definitely safe for consumers and the environment.

At an earlier meeting in March 2000, in Larnaca, Cyprus, the Greens passed the following resolution: "The only way to stop genetic engineering corporations from appropriating nature is to stop patents on living organisms." An important conference was held in Paris in February 1999, where the Greens dealt, in general, with GMOs and biotechnology in stronger terms. Their position can be summarized as follows.

(a) There are potential direct and indirect risks to human health from transgenic food products,

(b) Unacceptable risks to environmental well-being arise from the commercial dissemination of GMOs,

(c) There is a world-wide threat to biological diversity and to organic production from genetically engineered crops and animals,

(d) The European community's regulatory regime fails to respond to citizen's concerns,

(e) The U.S. sabotaged international efforts in Cartagena, Colombia, to agree to a Biosafety protocol to the CBD (Convention of Biological Diversity), and

(f) Monsanto has announced that by the year 2000, 100% of U.S. Soya beans (60 million acres) will be genetically modified.

The European Green parties demanded a five-year moratorium on the genetic manipulation of crops and animals used for foodstuffs, comprehensive labelling of all existing genetically-modified products, a stop on all GM-imports from the U.S.A., a ban on the patenting of life forms, and an independent and transparent re-assessment of the risks of all GMO releases to human health and to the environment.

The center of anti-GMO and anti-biotechnology movements appears to be concentrated in a few countries of western Europe, particularly in Great Britain, France, Italy, Belgium and Germany. For instance, the Minister of Agriculture in Italy, Alfonso Pecoraro Scanio, is a Green Party member who has banned all field trials involving genetically modified organisms (GMOs). He is a longtime critic of transgenic crops and claims that GMOs pose a threat to human health and the environment. In July 2000, he informed all 23 Ag-Biotech research institutions under the ministry that future research funding will be available only to those who would renounce field trials of GMOs (*Science*, 15 December 2000, p. 2046). Later, he extended the rule retroactively to all projects approved since 1996. Furthermore, any research proposal involving GMOs has not received funding or simply postponed until it is too late for planting. The protests of numerous Italian scientists including the Nobel laureate Renato Dulbecco have been ignored by the Agriculture Ministry.

Although the anti-GM groups in Great Britain were slow at first in their reaction to the introduction of GM technology in crops, they have become stronger in recent years, currently occupying a significant position in British Labor Party politics. Oddly enough, they share their platform with Prince Charles, an unlikely candidate to oppose biotechnology. In an article which appeared in the Daily Mail, he posed a number of questions about the safety and desirability of GM crops and their potential use to feed the world's growing population. The Prince quoted African sources, saying that GM crops will destroy diversity, local knowledge and sustainable agricultural systems, and will undermine the independence of local farmers. He suggested further that the question can be resolved only through independent research over a long period. Rigorous testing and field trials are necessary before any GM seeds are released to farmers. Labelling foods is another safeguard. According to the Prince, the argument that GM foods are necessary to solve the world's hunger problem is simply "emotional blackmail." It has been reported that the British Government was not happy with the statements of Prince Charles, which have once again reignited the national debate on the safety of GM foods.

## Blair Statement

British Prime Minister, Tony Blair, was initially a strong supporter of GM foods and stated in 1999 that he ate them himself. However, by February 2000 he had modified his position, stating that there is potential for harm to the environment as well as to the public. In an article in *The Independent*, dated 27 February 2000, Blair wrote: "...the protection of the public and the environment is, and will remain, the Government's over-riding priority... But there is no doubt, either, that this new technology could bring benefits for mankind. Some of the benefits from biotechnology are already being seen in related areas such as the production of life-saving medicines... GM crops, too, have the potential for good — helping feed the hungry by increasing yields, enabling new strains of crops to be grown in hostile conditions, or which are resistant to pests and disease." Blair promised that no GM food will be put on the market without going through the most rigorous safety assessments in the world.

## Valuable scientific research being destroyed

Among the many violent protests that were justified in fighting the planting of GM crops in Britain, genetically modified crops were destroyed in two demonstration plots of Monsanto's experimental "Roundup-Ready" sugar beet at Vine Farm, Wendy, near Royston. The main event was organized by the Royal Agricultural Society of England, the Home Grown Cereals Authority, the ADAS farm advisory company and the east of England Agricultural Society. It is the main national show in Great Britain where farmers can see the latest cereals, oilseed rape, and sugar beet varieties growing in demonstration plots. Other demonstration plots, containing non-GM varieties, were also destroyed. One of the protesters said: "GM crops are bad for the Third World and bad for the environment. We have only destroyed commercial promotional crops..." (*Telegraph*, 17 June 1999).

A leading British scientist, John Macleod, the Director of the National Institute of Agricultural Botany, described attacks on trial crops as "acts of mindless vandalism." He added: "...he was furious to see valuable scientific research being destroyed needlessly. Without this data and research it is impossible to reach a balanced view on GM crops... Our independent research programmes will contribute valuable technical data to help the authorities and others to come to sound decisions on the potential benefits and possible risks of these new technologies. It is essential that these trials are conducted to produce sound, objective information."

## British Medical Association

In a recent statement, the British Medical Association listed the following points of concern.

(a) Data are urgently needed about the effect of GM on the chemical composition of food, and its safety,

(b) Adverse effects resulting from the release of GMOs into the environment will be irreversible,

(c) GM foods are not equivalent to their conventional counterparts in safety (contrary to the assumption made for U.S. regulation of biotech foods),

(d) There should be a moratorium on the commercial planting of GM crops until the effects of GMOs on the environment and wildlife are thoroughly studied, and

(e) There should be a ban on the use of antibiotic resistance marker genes in GM food, as the risk to human health from antibiotic resistance developing in micro-organisms is one of the major public health threats that will be faced in the 21st century.

## Genetic weapons

In an earlier statement issued in 1997, Dr. Vivienne Nathanson, Head of science and ethics for the BMA, cautioned that gene therapy could be misused for developing "terrifying genetic weapons" that target and destroy ethnic groups. The same technique that is used for gene therapy could be used to achieve dangerous goals. Viruses or chemical compounds could become potential weapons in biological warfare with a new twist because of molecular biology. They could kill, injure people or make them defective or infertile. Such compounds could be delivered as a gas or spray, or put into the water supply. International cooperation is essential to ban such weapons. However, if we wait too long, vested interests would oppose their banning because of economic, political or other reasons.

# CULTURE AND BIODIVERSITY

# Culture and Biodiversity

Biodiversity represents the total biological wealth of planet earth. It encompasses all life — macro- and microbiological, aquatic, terrestrial or in any other habitat, including all plant, animal and human life. Although the practical and commercial value of biodiversity has been known from ancient times, the full impact of its worth as a national treasure has become obvious in recent times mainly because of the development of biotechnology as a distinct discipline. The immense value of biodiversity is derived from its many indispensible uses; for instance, crop biodiversity is essential for food production, lumber and wood pulp are used in a variety of ways, several plant species as well as some animal species yield medicinal products, our dairy and meat products are derived from animal biodiversity as are fish and poultry, and various secondary products derived from animals and plants have great commercial value. These are only a few examples of how dependent our lives are on biodiversity.

## Culture and technology

Life, land and community are often in balance, that is to say, they maintain a certain balance and consistency in their interrelationships. That balance is crucial for the maintenance of biodiversity, yet science and technology do not as a rule concern themselves with the values of local knowledge and traditions. Economists and scientists have conspired, though inadvertently, to moralize globalization. The terminology of "sustainable development" and "partnership" as well as of biodiversity conservation is alien to many traditional, indigenous and local communities, who equate technology and globalization to market strategies and economic exploitation as well as loss of biodiversity. Science and industry must respect local diversity and indigenous values, and that includes a recognition of the "sacred balance" between life, land and society. Many indigenous communities consider that balance in the same manner as the native Hawaiians — the *lokahi* (unity) — the nurturing, supportive and harmonious relations that link land, the gods, humans and the forces of nature.

It has been argued that the concepts underlying such words as "nature," "biodiversity" and "sustainability" were understood by the indigenous communities since the dawn of civilization. These were not invented by western science. Traditional and local communities have practiced the conservation of plants, animals and ecosystems for millenia. For instance, fire use was part of an extremely sophisticated system that shaped the balance of native flora and wildlife. They have molded and conserved crops and shaped the environment to such an extent that it is often not possible to distinguish culture from nature. Their efforts have long contributed to our survival and well-being.

The Kayapo Indians of Brazil used detailed knowledge of soil fertility, micro-climate and plant varieties to skillfully plant and transplant useful non-domesticated species into wooded concentrations of useful plants (or forest islands called "apete"). However, botanists and ecologists have considered them to be products of "nature," failing to recognize them as the products of Indian efforts. Similarly, the Ontario resource managers have failed to recognize the anthropogenic wild rice (*manomin*) fields of the Ojibway.

## The influence of religion

Today's biodiversity is the ultimate result of the conservation and nurturing efforts of indigenous peoples over several millenia. Many species of plants and animals have acquired religious significance over millenia. Deities in plant and animal forms have been objects of prayer and worship for thousands of years. Such an association promotes reverence and conservation towards all life. The *Bodhi* tree (*Ficus* spp.) is worshipped by Buddhists because Buddha was believed to have sat under that tree while meditating. The *Tulsi* plant (*Ocimum* spp.) is worshipped by Hindus daily with their morning prayers. Many believe that it possesses healing powers and its presence in the backyard contributes to the well-being of the family.

## Sherpas

The Sherpas, who are mountain people of Tibetan origin, have an intimate knowledge of their native Khumbu homeland in the Mount Everest area of the Nepalese Himalayas. They recognize several field sites, pastures, forest areas, tributary valleys, mountainsides and settlements which are not usually seen on any tourist maps. Their religious values and reverence for sacred places have played an important role in conserving land-use and natural resource management. Their traditional ideas and conservation practices have shaped the Conservation program of the Mount Everest National Park. Sherpas believe that all wildlife and plants have consciousness and spirits and, following Buddhist tradition, they have adopted a non-violent lifestyle. They refuse to harm even insects and avoided hunting for either subsistence or sport. Their attitude created a wildlife sanctuary throughout the region long before the establishment of a national park. Sherpas' respect for wildlife is mainly responsible for the abundance, in some regions, of musk deer, local species of pheasants and Himalayan tahr, and other species that are rare or endangered in the high mountain areas of Nepal.

Sacred forests and trees are common throughout the region where cutting down trees is forbidden by the lamas. Sherpas believe that certain mythical spirits, known as *lu*, live in Khumbu inhabiting springs, boulders, trees, shrines and houses. These spirits often influence the health and luck of nearby families and must be protected. *Lu* trees within villages are easily recognized because of the specially built shrines at the foot of the trees. The conservation of sacred forests is in sharp contrast to the region's non-sacred forests, which are the main source for timber and fuel-wood. The beneficial result of centuries of protecting sacred forests has led to the survival and abundance of many old trees of juniper, fir, birch and rhododendron in the Himalayan region. This is one of the best examples of biodiversity conservation that has resulted from the tradition of sacred groves and forests.

# The Peasant View of the Forest and Its Impact on Biodiversity

The idea of "extractive reserves" was promoted by the late Chico Mendes in the Brazilian Amazon to bridge the human-nature dichotomy by combining the livelihood of indigenous people with activities that sustain the Amazon forest ecosystem. An "extractive reserve" is a land that is under public ownership, but certain indigenous people have the right to live and work there.

Certain political units of modern India have been traditionally recognized as "ecological entities" long before the political concept took shape. One such region is the Jharkhand region (meaning forest area) which is located at an altitude of 500 to 1000 meters above sea level. The hilly region is noted for its own unique forestry, agriculture and irrigation technologies. The inhabitants have been called by various names which reflect the ecology of the region. These are:

| | |
|---|---|
| *adivasi* | (the indigenous people) |
| *girijan* | (mountain dwellers) |
| *vanajati* | (forest dwellers) |
| *paharia* | (hill dwellers) |

The region represents a curious mix of peasant knowledge and scientific resource management. Peasant knowledge and outlook encompasses a detailed knowledge of individual plant and animal species as well as their medicinal use, combined with crop diversity in a holistic way for practical purposes. Yet, it also places that entire outlook within the cosmological context in its proper place and time.

Although the indigenous people of the region are settled agriculturalists, they are fully conscious of the forest wanderings of their recent ancestors. There is a depth of interdependence between forests and *adivasi* peasants, whose struggle to survive led to a unique lifestyle and cultural as well as political outlook. The resulting moral and ethical code is based on utilitarian views, but also on moral reciprocity, restraint and prudence.

*Ecological prudence* in this context is defined as a "system of thinking and practice that maintains the natural process of renewal, the notion of sacredness and the compulsion of utility in day-to-day life, worked together as a moral constraint against destructive resource use." (Laird 1999, p. 386).

Biodiversity is maintained by a delicate balance between society and the natural and the supernatural world. Social order is maintained by a combination of taboos, protection from spirits, drought, floods and crop failure, as well as a fear of retribution from the spirits. The deities are appeased to cause rain and protect livestock and human life.

The "ecological cosmovision" of a community is believed to encompass ever widening and intersecting areas of human, natural and supernatural spheres, the material basis for human activity being provided by the "natural" sphere. For the Jharkhand *adivasis*, a delicate balance rests between the three spheres, reinforcing the system as a whole. It is described as a dialectical development involving all three components.

The folklore, music, songs and poetry of the region reinforce the view that the commercial use of new forests is quite distinct in character from the thought behind the creation that goes far beyond commerce; it is "divine." The forest is communal property. It is protected by the "sacred" nature of certain species of plants and animals, ponds, mountains, meadows and forests. Certain rules are strictly followed to promote conservation, such as not fishing the entire stock in a pool and protecting a pregnant deer during the hunt.

Deities play another role in protecting forest ecology as certain patches of original forest are maintained as *Sarna*, where the main deities of the clan are kept. This concept is said to date back in origin to pre-Vedic times (about 3000 B.C.). Some groves are known as *Sharana* forests, meaning sanctuary. Other protected areas include graveyards, dancing or assembly grounds, water resources, and occasionally mountains and watersheds. A *Sarna* is different from the modern national park system, which is protected from human use. Ironically, the British in India created "reserved" or "protected" forests (under the mistaken notion of a *Sarna*) to keep them beyond the reach of the local communities. The *adivasi* identity is intricately linked with the culture of sacred groves of *Sarna*, which, as discussed earlier, occupy a central part in their ecological cosmovision. Human, natural and super-natural spheres seem to coexist in equivalence and conformity. The general ecology of the area is viewed as a continuum, ranging from farms to forests, pastures and settlements. In contrast to the mainstream ecological movement in several countries, the *adivasi* view avoids the trap of falling into the culture/nature divide, which is clearly a more meaningful and holistic approach to the conservation of biodiversity.

## Shifting demographics and biodiversity loss

The definition of the term "indigenous" itself is fraught with many political and social dilemmas. Demographic patterns have been constantly shifting, with immigrants flowing into tribal areas in search of trade in forest products and development. These changes have, in turn, adversely impacted on the biodiversity of certain regions. The non-tribal immigrants, who are mostly Hindus, do not share the reverence and the complex belief system that has played a large part in conserving the forests and wildlife. Larger economic forces and commercialization as well as mechanization are threatening to destroy the delicate ecological balance that has existed for centuries. Joint Forest Management (JFM) programs, involving co-operative agreements between village communities and the local Forest Department, utilize both local knowledge as well as the larger interests of development. Under these programs, 1–2 million hectares of forest land is being protected. The programs of JFM generally include a focus on micro-planning including planting selected species chosen by the local people, and utilization of traditional peasant practices such as appropriate gathering practices, grazing, lopping and firing, and so on, which were initially denigrated.

In several instances, the major threat to forest land came from commercial interests such as tourism and bauxite mining in Borra in the state of Andhra Pradesh, power plants in Sambalpur, and paper mills in Gujarat. In some instances, the villagers' views were taken into account in the selection of forest areas to protect and the choice of species to plant in those areas. Indigenous knowledge of local biodiversity is utilized. In one village in Bastar, the local women identified 51 species of plants which were useful to them in various ways. In several villages in the state of Andhra Pradesh, the villagers asked for cash crops and new species to be planted (e.g. coffee). They did not ask for the traditional species of trees in their area as they expected them to regenerate and grow in due course. In other cases, as in West Bengal and Orissa, women asked for sal (*Shorea robusta*) plants because of the commercial value of their leaves as dinner plates, which is the main source of income for those women. Decisions are thus influenced by the economic, spiritual or other needs of the local communities as well as the government regulators. It is the dynamic nature of the indigenous peoples' initiative that is impressive.

# Sacred Groves and Biodiversity

Sacred groves are kept intact because of their sacred nature, and the strong cultural and spiritual values that are traditionally associated with them. Examples include the "dragon hills" of Yunan Province in China, and others in India, the Ivory Coast, Benin and Ghana. Sacred groves are specific forest areas that are believed to possess powers beyond those of human beings. They are often associated with mythological characters and mighty spirits. Rivers and streams flowing in those areas confer healing powers upon the priests. These areas are treated with special respect and access to the most sacred forests is restricted by traditional customs and taboos. Hunting, wood-chopping, cultivation and other similar activities are strictly prohibited in the Holy Hills of China.

## India

In south India, trees have played an important part in the funerary rituals of certain castes. Interestingly, they may be associated with some benefits for the souls of some higher castes while resulting in misfortunes for those of lower castes, especially "untouchables." Tribal people who inhabit the forest belts throughout India have preserved certain religious and cultural practices that are associated with areas of forest land, which, in turn, helped to establish a pro-conservation behavioral pattern. The "sacred" parts of forests have been protected for many years and have become the ecological niches of rare and medicinal species of plants.

The "sacred" groves in the Western Ghats of India are surrounded by highly heterogeneous populations, varying in their caste diversity and occupations. Many depend on forest resources for food, medicine, fodder and fuel. The forest provides the basic medical needs of the surrounding communities as well as the tribal people. Even the simplest western medicine is unaffordable and beyond their reach.

Among these tribal communities are specialist groups who possess valuable knowledge of medicinal herbs, the healing properties of roots, leaves, fruit and other parts of various species, and where to fish, hunt and capture wildlife. And there are *pujaris* or priests in the surrounding communities, who have defined the relationship between the deities and the community traditionally. They have assumed the responsibility for the implementation of local rules and regulations, based on religious traditions, leading to the conservation of regional plant diversity.

"Sacred Groves" are usually quite small in size. The following is a classification of the forty groves studied in the area.

| Number of groves | Size |
| --- | --- |
| 30 | <1 hectare |
| 4 | 1–2 hectares |
| 4 | 2–3 hectares |
| 1 | 4 hectares |
| 1 | 8 hectares |

Some of the groves are semi-evergreen; others are evergreen. The biodiversity of the region is indicated by the fact that, in the Khandala Hills, 800 plant species were reported, of which 162 are trees and 82 are shrubs. In a study of 15 groves, the total number of species of trees and shrubs recorded is 223. The number of trees/shrubs recorded per grove varied from 10 to 86. The groves represent the least disturbed ecological niches in the region. The numbers mentioned do not include the large number of ground flora and annuals that appear in the rainy monsoon season.

Several temples, which are presided over by the local Brahmin priests, are located in each of the villages in the region. There are also simple forest deities, located in the depth of the forest grove, on a hill or near a water source. The surrounding area is sanctified, resulting in forest conservation. Traditional rituals include animal sacrifices, controlling weather and predicting other future events. Such tasks are the responsibility of a specific tribe or sub-caste in each region.

The "ownership" of the groves is varied, ranging from private to public ownership. The following is the ownership classification of the forty groves studied.

| Number of Groves | Ownership |
|---|---|
| 15 | Village *Panchayat* (council) "Commons" |
| 13 | Reserve Forest Land (owned by the "deity") |
| 11 | Private ownership |
| 1 | Temple Trust |

Some limited use of forest products is permitted by the "deity" (permission obtained following a ritual), including the collection of fallen dry wood, fruit, and honey and sap for making liquor. In most groves, only fallen wood is permitted to be gathered from the forest floor. Cutting down trees is rarely permitted. There is a taboo against allowing women (especially menstruating women) to enter the "sacred Grove." Men rarely collect wood as it is considered a woman's task. Much of the fallen dry wood is not collected. However, the taboos and restrictions vary from village to village.

Local folklore is replete with stories of how certain supernatural incidents have been associated with any attempt to destroy the groves. The usual consequence is the occurrence of illness after someone has cut down a tree in the grove. Stories of the sudden occurrence of blindness and sickness when women entered the grove are told frequently in the region.

It appears then that although the "sacred groves" were not established primarily for species preservation, they have come to play an important role in forest conservation. Maintenance of biodiversity is a byproduct of this highly complex phenomenon, which is a mix of religious and folk rituals as well as a form of socio-political organization of the society. They seem to be based on a complex range of deeper emotions, from admiration of a deity to fear and love, the ultimate result being the conservation of biodiversity. However, modern economic pressures seem to be encroaching upon these groves as some of the private owners are selling them to build tourist hotels.

# ISSUES IN DNA TECHNOLOGY

# Who Owns the DNA?

First, let me acknowledge that the patenting of life forms or segments of cellular tissues such as the DNA is repugnant to many people, who object on the basis of religious, ethical and moral grounds. However, we must at least understand the process by which intellectual property rights of life forms, living tissues and cells, are claimed because it is playing a greater role in the way research is supported and organized and is having a huge impact on agriculture and medicine.

The issues we have to confront in claiming Intellectual Property Rights (IPR) *en masse* are somewhat different from the problems involved in patenting individual genes. Although both involve priority claims, individual patents may be recognized within a broad framework of IPRs. In other words, patents are the tools that are generally, though not always, employed in recognizing and protecting IPR.

Numerous inventions in biology and biotechnology have come forth in recent years that touch upon important aspects of medicine and agriculture. The problem is further complicated by international competition and the disparity between nations in enacting appropriate legislation and resources to implement IPR.

## Some global issues of IPR in biotechnology

There is no uniformity in the standards nor in the degree of enforcement of IPR in different countries. Any IPR system should satisfy certain requirements such as: (a) it should be flexible enough to incorporate rapidly evolving and changing technologies, (b) it should maintain uniformity in recognizing all forms of intellectual property, (c) it should be supported by an efficient public administration, and (d) an appropriate judicial system should exist to enforce individual rights.

An IPR system is necessary to inspire confidence in foreign investment. It would promote technology transfer at different levels. Increasingly, research collaboration between academia, industry and government is creating new situations which would not be possible without IPR agreements. Basically, developing countries are being transformed from a traditional climate of collective and public good into individual rights and private benefits. The conservation of biodiversity and protection of rare species must go hand in hand with this transformation.

It is estimated that 74% of the plant-based prescription drugs on the market today have the same use in western medicine as originally practiced by native healers. But no royalty or compensation has been paid to native communities for their intellectual contribution. However, Shaman Pharmaceuticals has pioneered the direct utilization of indigenous knowledge in the drug discovery process. They have pioneered the practice of compensation in advance, even before drug exploration or development is initiated.

## Issues in patenting DNA

Is DNA a chemical compound? The central part of biotechnology today is the DNA structure, variations and functions. For patents, DNA sequences are treated as large chemical compounds. This approach is advantageous to those who seek to patent new DNA sequences, but not to those who are interested in elucidating biological functions and disease causation by specified genes. In recognizing patents, U.S. courts have generally followed the rules that were applied for chemical inventions. Thus, according to the definition, DNA sequences are regarded as large chemical compounds and are patentable as compositions of matter under the same principles that were drafted previously for small molecules. Under this definition, courts have circumvented the sacred rule against patenting "products of nature," and they were able to recognize claims covering "purified and isolated" DNA sequences, recombinant vectors and host cells that include these sequences. This recognition is based on the precedent involving other approved patents for purified versions of adrenaline, prostaglandins, vitamin B-12 and acetylsalicylic acid. Instead of focusing attention on the method of making a new chemical, the courts have concentrated on the structural and functional differences between the claimed compound and other compounds in the "prior art."

In summary, the patentability of a new DNA sequence depends not on the inventive skill required to obtain the sequence, but rather on the absence of prior disclosures of similar DNA molecules. It is clear that what is patentable is not necessarily considered to be an outstanding achievement by the scientific community.

## The human genome

The Human Genome Project is resulting in a number of new patents. Sequencing the human DNA involves about three billion nucleotides. Goals of the project for the five-year plan (1998–2003) include a complete human genome* sequence by the end of 2003 (two years earlier than originally planned).†

Minute changes in individual genomes can lead to deadly diseases. Knowledge of single nucleotide polymorphisms (SNPs) is providing valuable clues to differentiate between healthy and diseased tissues. Pharmaceutical companies are vigorously competing to protect this knowledge by filing patent claims.

## Who owns the human genome?

The principle of respect for autonomy is one of our most valued human rights, and hence it follows that a person's autonomy requires informed consent before sampling a person's DNA. Further, the individual's permission is required before the results are made public. When some

---

*The genome of an organism is its complete set of chromosomes, containing all of its genes and associated DNA.
†It was actually completed in 2001.

individuals are identified as carriers of specific diseases at some future date, this inevitably brings forth questions or dilemmas which confront those who carry the risk, as well as others, such as their potential employers, insurance companies and so on.

These factors indicate that genetic knowledge of ourselves is also an intellectual property which must be protected. It becomes even more complicated as it involves questions of personal ownership and privacy such as: Who owns the human genome or DNA? Is it collectively owned by society? In that case, how can we justify individual patent claims or treatments that are derived from human DNA? Is it fair to charge me for a drug that is based on my DNA?

Let us imagine that a pharmaceutical company agrees to pay a recurring annual fee, a certain percentage of its profits, after deducting its cost of drug development and production. Will that satisfy all of the parties concerned?

There are other related questions: Where should we draw the line between private ownership and public access for the common good? If each individual claims intellectual property of his own genome, will it not hinder research for the common good of society?

Is it right to claim private patent protection for research that was supported by public funds? Large scale DNA databases are being stored privately and publicly for future use. To what end will they be used, and who will make the decisions regarding their future use? Who will have the right to utilize this kind of intellectual property in the future?

Perhaps a fair and equitable arrangement for sharing benefits should be agreed before such research commences. Even population DNA sampling should be undertaken only after we have resolved these dilemmas.

## New issues

New ethical and legal problems are cropping up, such as, modifying or altering genes, thereby creating new varieties or species. Once again, we face the same issues stated above, with the same ethical and moral dilemmas. Who will have the ownership of the genome, and who will make the decisions to alter the genome and to what end? Are we going to agree on which direction we are to proceed?

Does ownership give the right to alter a gene or the entire genome? In somatic cell gene therapy, we attempt to correct a disease-causing gene or replace that gene in the somatic tissue. It can only be carried out after obtaining the informed consent of the patient and his immediate family. Since it is only somatic tissue, alterations in the genome are not passed on to future generations.

On the other hand, in germ line therapy (which is yet to be attempted), any alterations of the genome will be long-lasting, persisting through several generations. It could, in the long run, alter the genetic composition of a population. Therefore, alterations should be attempted only after discussions to ensure that the expected benefits to society outweigh any possible adverse consequences.

# DNA Research, Gene Therapy and Cloning

The practical applications of DNA research may be more appropriately called the useful applications of various recombinant forms of the basic molecule which carries genetic information, DNA or deoxyribonucleic acid. This is present in the body cells of all living organisms. Long molecules of DNA are tightly packed in thread-like structures (called chromosomes) in each cell. There are 23 pairs of chromosomes in each cell. All of our physical, mental and behavioral characteristics are mostly determined or influenced in some way by the genetic messages coded in the DNA molecule. There are approximately 50,000 to 100,000 human genes which account for what we are, how we look, and how we behave or perform in our lives. A genetic disease is the result when the structure or function of the corresponding gene is impaired. This happens spontaneously in nature in each generation, but it can be greatly enhanced by atomic radiation, chemical contamination, or some other reason.

## Gene therapy

The goal of human gene therapy is to insert into our body cells, corrective genes (pieces of DNA), either to repair defective DNA or to replace a defective gene entirely. Before we can do so, genes causing specific defects and their exact location on a chromosome must be identified first. This is the function of the human "genome" project, an international program which is designed to produce complete maps of all of our genes in the next few years. DNA research has already resulted in many useful applications which are beneficial to us. The pharmaceutical industry had been attempting to produce synthetic insulin, growth promoting substances, and synthetic allergens, as well as devising methods for treating cancers (such as anti-sense techniques), and for improving the capacity of individuals to metabolize medical drugs effectively. One important field is the creation of "transgenes," or transfer of genes between different species which makes it possible to produce large quantities of various enzymes and hormones (in other species) which are beneficial to mankind.

## Cloning

Clones of different animal species with recombinant DNA can be created for producing large quantities of insulin or some similar product. Cloning of specific organs for surgical or medical transplantation will become a common practice in the future. On the other hand, cloning of whole human beings does not appear to be an immediate necessity even though it may be technically feasible. Clones of crop plants and fruit trees have been useful to humanity for many years. In the future, clones of animal species will greatly benefit agriculture and food production. Cloning of whole human beings for specialized purposes may be attempted soon by governments or private corporations.

# DNA Technology in Forensic Medicine

Each year, the Federal Government of the U.S. spends several hundreds of millions of dollars on basic research of all kinds. At first glance, many of these research projects may seem not to have any practical application or may even seem frivolous. But one can never predict what basic research can produce. Past experience tells us that several practical applications have emerged from this kind of approach. One example is the discovery of the first antibiotic, penicillin, which was discovered quite accidentally during what seemed to be fairly routine and basic studies of various molds. When a petri dish containing a mold was accidentially left overnight without a cover, it was found later to be contaminated with bacteria which are, of course, always present in the air we breath. What was most interesting to the scientist, Dr. Alexander Fleming, was the fact that bacterial growth was sparse or absent in the vicinity of the mold. It was thus deduced that some substance(s) secreted by the mold was detrimental to bacterial growth and survival. As we now know, the use of penicillin and other antibiotics has saved several millions of lives.

The above example is probably the best reason why we spend a great deal of money on basic research. Much of the early work on atomic structure and rocket technology started as benign basic research in physics, but ultimately resulted in powerful ballistic missiles with nuclear warheads. We cannot always foresee what basic research might bring about.

# DNA technology

The recombinant DNA technology which gave birth to modern forensic applications of "DNA fingerprinting" has its roots in early research into the structure and function of living cells. DNA or deoxyribonucleic acid carries the coded messages which ultimately determine what we are, what we look like and how we behave. Almost all characteristics, physical as well as mental, are influenced by the DNA which is influenced to some extent by environmental, dietary or other factors. Thus, although a person may carry genes for diabetes, the degree of manifestation of the disease depends to some extent on the diet of the individual.

Every living being (including all animals, humans, plants and microorganisms) is made up of billions of microscopic cells, and each cell contains a certain number of chromosomes. Each chromosome is a thread-like structure which embodies the DNA. The number of chromosomes per cell is constant for the species, that is to say, all human cells normally have 23 pairs of chromosomes. Some minor deviations from this number are occasionally noticed; some are benign with no pathological consequences whereas other deviations or multiple numbers are often associated with specific diseases, such as "mongolism" (or what used to be called "mongolism") and various types of cancer tumors, etc.

The chemical structure of DNA represents the coded message of our hereditary information. The code is commonly represented by the arrangement of four letters, each denoting a nucleotide: A (adenine), C (cytosine), G (guanine), and T (thymine). A nucleotide is a unit of nucleic acid. Various combinations of three nucleotides indicate the codes for the amino acids which make up our proteins. The genetic material in a cell usually consists of about 50,000 to 100,000 genes, each gene containing normally about 1,000 to 2 million nucleotides. The total genetic make-up of a human being is called the "human genome."

# DNA polymorphism

I will not go into many technical details, but it is important to realize that advances in DNA technology in the 1970s paved the way for the detection of variation in specific DNA sequences. The term "polymorphism" has been used to describe such variation because it involves the existence of two or more forms of a certain characteristic. When a sufficient number of DNA regions are analyzed, it is easy to show that much variability exists between individuals and that the probability of a chance match between two persons is so low that DNA typing can be used for absolute identification.

## DNA fingerprinting

The term, "DNA fingerprint," was first introduced by Dr. Alex Jeffreys in England in 1985, demonstrating its potential as a powerful tool in forensic medicine. Forensic DNA typing was initiated in the United States in 1986 by commercial laboratories and in 1988 by the Federal Bureau of Investigation (FBI). In 1992, a report on the forensic aspects of DNA technology was issued by a committee of the National Research Council of the National Academy of Sciences in Washington, D.C. It was funded by the Federal Bureau of Investigation and was strongly supportive of the use of DNA fingerprinting in forensic medicine. The Chairman of the Committee, Dr. Victor A. McKusick of Johns Hopkins

### DNA Content of Biological Samples

| Type of sample | Amount of DNA |
| --- | --- |
| Blood | 20,000–40,000 ng/ml |
|     stain 1 cm$^2$ area | ca. 200 ng |
|     stain 1 mm area | ca. 2 ng |
| Semen | 150,000–300,000 ng/ml |
|     Postcoital vaginal swab | 0–3,000 ng |
| Hair | |
|     plucked | 1–750 ng/hair |
|     shed | 1–12 ng/hair |
| Saliva | 1,000–10,000 ng/ml |
| Urine | 1–20 ng/ml |

The amount of DNA is given in nanograms (or ng). One ng = one-billioneth of a gram ($10^{-9}$ g).

University stated: "DNA typing for personal identification is a powerful tool for criminal investigation and justice." The preceding table is based upon this report. It shows that the very small amounts of DNA which can be found in forensic samples are adequate for testing and DNA typing.

## Concerns about DNA fingerprinting

It is precisely because of the power of this tool that DNA fingerprinting needs to be regulated and monitored very carefully. Private laboratories and businesses may attempt to "cash in" on this new technology. Indeed, some have already done so. Unless quality control is exercised, it can fall into disrepute and lose credibility. There are several possible sources of error. The report of the National Academy of Sciences/National Research Council Committee on DNA Technology in Forensic Medicine made several recommendations for maintaining high standards:

(a) A new DNA typing method or a modification of the existing method must be rigorously characterized in both research and forensic settings to determine the circumstances under which it will yield reliable results.

(b) The following rules should be observed to maintain the highest scientific standards in DNA testing for forensic science.

   (i) Each DNA typing procedure must be completely described in a detailed written laboratory report.

   (ii) Each DNA typing procedure requires objective and quantitative rules for identifying a sample.

   (iii) A precise and uniform matching procedure must be observed for declaring whether two samples match. (I will not go into the technical aspects of this procedure except to say that it is not simple. The central question to be answered is this: "What is the probability that a person picked at random would match the evidence sample in DNA patterns?" Obviously one needs to have a detailed knowledge of the DNA patterns of the right population of which the suspect was a member. Two other points need to be noted. There has been much discussion among scientists as to the number of alleles (variant genes at specific loci on the chromosomes) as well as the number of loci (often four) on a chromosome which must be tested for a reliable result. Secondly, it should be noted that the frequencies of the various alleles for the DNA polymorphisms may vary greatly between populations.)

(iv) Potential artifacts should be identified by empirical testing and internal controls should be tested for the occurrence of artifacts.

 (v) The limits of DNA typing (especially when the sample is small and a mixture from different sources are used) as well as the possibility of contamination should be clearly understood.

(vi) Before a new DNA typing procedure can be implemented, it must be thoroughly tested for its scientific basis as well as for its practical applicability.

(c) As there are different methods of DNA typing, the Committee made a number of technical recommendations about the reliability of each method. One method is called the "RFLP method" in which the DNA is subjected to controlled fragmentation with restriction enzymes, and another method, the PCR for amplifying DNA, involves the use of the polymerase chain reaction (PCR) and allows a million more copies of a short region of DNA to be made. It is a method of DNA amplification which involves the genetic information in the amplified product.

(d) The Committee recommended the establishment of a National Committee on Forensic DNA Typing to provide expert advice on scientific and technical issues.

Under certain circumstances, DNA matching evidence can play a crucial role in pointing to an individual as the most likely suspect in a crime. However, one presumes that an individual is innocent until proven guilty in the normal judicial process. Where there are no eye witnesses to a crime and no murder weapon found, purely circumstantial evidence may point to one individual. In such cases, if the DNA pattern of the chief suspect matches the blood samples found on or near the murdered victim, then it would be hard to avoid a guilty verdict. In other words, DNA evidence is so precise and the probability of other possibilities so negligible that DNA evidence can overwhelm any other argument by defense attorneys.

In such instances, science can make all the difference. After listening to the DNA evidence from an expert, the jury would be in no mood to be swayed by other arguments.

## Biomedical ethics

DNA technology should be subjected to the same ethical and moral standards as any other biomedical technology which involves questions of personal privacy. Some years ago, the presence of an extra Y chromosome in the cells of a male (XYY instead of the normal XY) was supposed to predispose these carriers to taller stature, aggressive behavior and violent crime. But as the initial surveys were performed in penal institutions, these early impressions were based on a study of biased samples of populations. Later studies showed that the XYY pattern can also be found in normal (non-penal) populations. While the impact of an extra Y chromosome on human growth and behavior is of great interest and merits continuing research, it would be ethically wrong to label all individuals with XYY as potential criminals. Similarly, it would be wrong to conclude that there are certain "criminal" DNA patterns. It is also important to remember that the DNA pattern of an individual is very private information of personal identity and must be treated as confidential. It should be revealed only under certain special circumstances for a specific legitimate purpose.

DNA technology introduces a much more sophisticated way of resolving cases of personal identity both in forensic medicine as well as in paternity cases.

All individuals have a right not to be wrongly convicted of a crime. For this reason, a high standard of proof is required before a person can be found guilty of a crime. The reliability of a test must be verified repeatedly for its accuracy and sensitivity. On the other hand, DNA evidence also provides a way of proving the innocence of potential suspects who might otherwise be falsely charged with a crime.*

---

*A report in December 2001 stated that about 100 inmates of Federal and State prisons in the U.S., who have been accused of serious crimes, have been found to be innocent by new DNA evidence. Misconduct by the prosecutors was implied in several of these cases.

## Budgetary costs

In a 1999 public hearing before a congressional committee, Mr. Louis Freeh, then Director of the FBI, requested that seven million dollars be allocated for the DNA testing laboratory of the FBI. However, with increasing crime rates each year, the projected costs will almost certainly increase exponentially. What is even more important in this context is that the U.S. Congress demands the appointment of a high level advisory committee to monitor the quality and progress of DNA testing procedures.

Past experience shows that every new technology has two sides. It can be very beneficial if properly used. On the other hand, it may turn out to be a disaster if abused. DNA technology is potentially a very powerful tool for not only forensic medicine, but in many other areas of biomedical research with direct practical applications. As a member of the U.S. National Institutes of Health's Recombinant DNA Advisory Committee, I have often evaluated cancer research proposals which involve another type of DNA technology, namely the introduction of cells with recombinant DNA, into cancer patients. This may eventually provide the solution to treating many forms of cancer.

# Public Perception of Genetic Engineering

The public at large is becoming increasingly aware of the issues involved in the application of genetic engineering, both its benefits and the potential risks involved. However, public perception depends on the sources and quality of information that is made available in a society. The term "public" includes individuals and groups of varying education, knowledge and perception. Information is disseminated through professional and public media. Consequently, public perception depends on the degree of accuracy of the information and the extent of technical knowledge that is transmitted through these media. Unfortunately, technical accuracy is sacrificed by much of the popular media to increase the extent of popularization. Errors of fact and judgement as well as personal bias often color the popular interpretation of scientific data. Other factors that influence public opinion include complex socio-political considerations and religious, moral or ethical dilemmas, as well as economic considerations. Risk perception plays a large part in accepting any new technology such as genetic engineering. Risk and safety are among the most frequently raised issues in relation to genetically engineered foods.

The following are some of the major factors which play a significant role in the public perception and ultimately the acceptance of genetic engineering.

## (a) Commercial success

Media coverage of genetically engineered crops and foods has heightened public awareness quite significantly in recent years. Press reports of the commercial success of new biotechnology often went hand-in-hand with dire warnings and possible risks. The commercialization of biotechnology and genetically engineered organisms first became a reality in 1980 when a genetically engineered bacterium (*Pseudomonas* spp.) developed by Ananda Chakrabarty for the purpose of breaking down crude oil, was recognized as a patentable subject in the U.S. The process did not involve creating life or new genes, as some have claimed, but only the reshuffling of already existing genes. In that landmark case of Diamond versus Chakrabarty, the U.S. Supreme Court ruled that a living organism is patentable if it involves a human-made invention. During the last twenty years, we have witnessed biotechnology's great impact on both agribusiness and the pharmaceutical industry, leading to its spectacular commercial success.

## (b) Risk perception

Genetically modified (GM) foods are accepted or rejected by the public with varying degrees of success, depending on the product, the geographic region and the socio-economic group. Generally speaking, medical applications are more readily accepted than those in agriculture. Some concerns regarding the use of GM foods are listed later. These include the persistance of genetically engineered organisms in the environment and their potential impact on the health of human populations. The contamination of crops, water supply, and farm animals is a major concern.

A fundamental concern centers around the feeling of "unnaturalness" which is associated with the creation of recombinant DNA and its applications in improving crops and various food products. Risk perception is greater when exposure to a risk is involuntary. It is then perceived as more threatening than when there is a choice over personal exposure (Sharlin, 1989). This perception is evident in relation to genetically engineered foods because of the inability to distinguish them from naturally occurring foods unless they are clearly labelled. What is involved is a feeling of a lack of control over one's food intake and its impact on our well-being.

## (c) Need

Public perception of GM foods is further influenced by the fact that it would appear to be a "superfluous" complexity of one's life, which is not warranted by daily needs. Perceptions involving the question whether this particular technology is necessary appear to be more important than perceived risk or ethical concerns. Responses to surveys often include a comment on the absence of any "need" to accept genetically engineered foods in our daily lives. On the other hand, genetically engineered drugs and other pharmaceutical products are considered necessary and hence are readily acceptable. Diagnostic and therapeutic medicinal products, even if produced by genetic engineering, are considered to be essential for a healthy life. The same sense of "need" does not, of course, exist for accepting genetically engineered foods. Perceptions of long-term effects were not found to be associated with either acceptance or rejection. Ethical concerns appear to be far more involved with genetic engineering than any other form of technology.

## (d) The Yuk Factor

The Yuk factor, which has been used to describe public reaction in a number of contexts including biotechnology, is almost always the initial public reaction to new biological discoveries. Several years ago, J. B. S. Haldane wrote: "Biology may not be taught to children seriously; that is to say, it may not be taught to them in connection with their own lives. Human physiology and digestion upset quite a number of our prejudices. The physiology of digestion, reproduction and excretion are indecent; the physiology of the brain is irreligious. On the other hand, chemistry, physics and certain branches of botany have no immediate bearing on conduct, and therefore they do not come into conflict with any deep-seated prejudices, and are taught in schools." (From *The Inequality of Man and Other Essays*, J. B. S. Haldane, Chatto & Windus, Ltd., London, 1932).

The feeling of "unnaturalness," expressed in the form of disgust, which is known to be associated with GM products, falls into the same category as Haldane's description of our reaction to certain branches of human physiology, i.e. the Yuk factor.

The Yuk factor is governed by instinct; it is primal and appears to be a basic law of human nature. It is expressed in varying degrees in reaction to different inventions and discoveries throughout history. Expanding knowledge and the passage of time are usually helpful in overcoming the Yuk factor.

## (e) Control

As mentioned earlier, risk is perceived to be higher in situations where an individual has no role in the decision-making process. This is indeed the case with GM foods. Control is an important characteristic of risk perception. Genetic engineering is perceived as a technology over which the individual has very little personal control. The relative roles of individual risk assessment and group assessment in making decisions regarding genetic engineering is not clear. The public appears to be capable of differentiating between abstract science and science directed at specific problems.

## (f) Ethical concerns

Public perception of genetic engineering and the eventual acceptance of GM foods appear to be influenced by ethical considerations. Ethical dilemmas are related to the intervention and modification of the naturally occurring genomes and the transfer of that technology. These concerns include:

(a) New technologies are in conflict with traditional beliefs and religious attitudes concerning the sanctity of life.
(b) New technologies may lead to the extinction of rare species, for instance, by genetic contamination.
(c) Patenting transgenic technology is not only contrary to many religious doctrines, but it may also lead to over-exploitation of some species and to the eventual reduction in biodiversity.
(d) Imported technologies, such as transgenic technology leading to GM foods, may compete unfairly with indigenous practices and industries.
(e) Copyright infringement and pirating of GM products is unethical.

(f) Developing countries may lack adequate safeguards and trained personnel to evaluate the adverse impact of GM seeds and foods.

(g) The application of DNA technology may lead to ethical dilemmas regarding our basic human rights, privacy and confidentiality of information, and genetic invasion of our genomes through the use of GM foods.

## Most frequently cited issues in genetic engineering

(Based on a study by Virginia Polytechnic; they are arranged in order of decreasing frequency)

(1) Should genetically engineered foods be labelled?

(2) Will the input from consumers be used in making decisions about the use of biotechnology products?

(3) Are current regulations providing adequate public safety?

(4) Will increased knowledge about biotechnology reduce concern?

(5) Will genetically engineered animals pose a health hazard?

(6) What is the best way to include the public in risk assessment?

(7) What are the ethical and moral concerns in creating engineered organisms?

(8) Are T.V. and other media accurately conveying biotechnology issues to the public?

(9) Are engineered foods less expensive and more nutritious?

(10) Are there acceptable risks for biotechnology and who decides?

(11) Are genetically engineered foods safe to eat?

(12) Is it ethical to develop genetic engineering for animals?

(13) How is the public protected from biotechnology products in developing countries with no regulatory system?

(14) Should animal welfare be a consideration in the genetic engineering of animals?

(15) What is the rationale for patenting engineered organisms?

(16) How does patenting impact upon the availability of scientific information?

(17) What are the risks to humans from large-scale releases of genetically engineered microorganisms?

(18) Will biotechnology contribute to the demise of small farms?

(19) How do we reconcile purely academic responsibilities with commercial exploitation?

(20) Will engineered microorganisms have any long-term effects on the environment?

(21) Can engineered crops become weeds?

(22) How can biotechnology products that are commercialized in industrial countries be made available to developing countries?

(23) Could engineered crops become a threat to native plants?

(24) Could engineered crops transfer new genes to wild relatives?

# The "Terminator" Gene

In recent years, there have been a great number of reports in the world press about the so-called "terminator" gene. However, there has been comparatively few in the United States, even though the "terminator" gene was the product of research there.

How do we account for this vast difference in impact? To understand this difference, we must first learn something about the nature of the "terminator" gene and its consequences to agriculture.

A patent for the "terminator" gene was jointly awarded in March, 1998 to the world's largest cotton-seed company (Delta & Pine Land Co.) and the U.S. Department of Agriculture, U.S. Patent Number 5,723,765. The gene has the unusual capacity to produce inviable seeds when introduced into the genome of certain crops. Consequently, there is no crop in the following generation.

Farmers who are accustomed to saving a certain portion of seeds every year to plant crops in the next season will not be able to do so because the seeds will not germinate into adult plants. They will be forced to buy fresh seeds each year to plant the next crop. Seed supply will be controlled by a few large corporations, which may very well achieve a global monopoly.

On the other hand, this situation will never prevail if the "terminator" gene is not introduced into the genomes of many vital crops. However, at present, there are plans to introduce the gene into rice, wheat, sorghum, soybeans and other major crops.

The president of Delta & Pineland Corporation, Mr. Murray Robinson, claimed that the new technology will have global implications. Delta & Pineland Company (in Scott, Mississippi) is the largest cotton seed company in the world, with annual sales of 183 million dollars in 1997. It controls almost three-quarters of the U.S. cotton seed market.

Monsanto (in St. Louis, Missouri) is a minor shareholder in Delta & Pineland Co. Monsanto is the world's second ranking agrochemical corporation. Its investment in seeds and agrochemicals over the past two years exceeded two billion dollars.

## Rural Advancement Foundation International (RAFI)

RAFI is an international non-governmental organization with headquarters in Canada, and a branch office in North Carolina. RAFI has termed the introduction of the new gene as "Terminator Technology" or the "Neutron Bomb of Agriculture."

A statement issued by RAFI described the situation as follows: "It is a global threat to farmers, biodiversity and food security. The seed-sterilizing technology threatens to eliminate the age-old right of farmers to save seed from their harvest, and it jeopardizes the food security of 1.4 billion people, as well as resource-poor farmers in the South who depend on farm-saved seed.

The developers of the technology say that it will be targeted for use primarily in the South as a means of preventing farmers from saving proprietary seeds marketed by American seed corporations. Delta & Pineland Co. and the U.S. Department of Agriculture have applied for patents on the Terminator technology in at least 78 countries.

If the Terminator technology is widely utilized, it will give the multinational seed and agrochemical industry an unprecedented and extremely dangerous capacity to control the world's food supply."

An additional danger is the possibility of infection of the fields of farmers, who either reject or cannot afford the technology, by pollen from crops carrying the "Terminator" gene. Their crops will not be affected that season, but some of the seeds may be found to be sterile when the farmers sow them next season, an unpleasant surprise!

Patenting the "Terminator" gene is part of a pattern of intellectual property rights which are increasingly claimed in western countries with profound consequences for the survival of many developing countries.

Global patents have a far-reaching deleterious impact on the economy of many poor countries. Several recent patents have encroached upon the sovereignty of developing countries, especially with respect to the ownership of biodiversity, which has traditionally belonged to these countries. Examples are basmati rice, neem tree products, turmeric and other species of medicinal or nutritional significance. Ironically, while these developments have been viewed with great alarm in developing countries, there is a total lack of concern and ignorance among the public in the developed world.

From an ethical and moral point of view, this situation raises a number of questions which must be answered sooner rather than later.

# Terminator Technology: A Re-Evaluation

There has been much concern and interest in the so-called terminator gene technology and its applications in recent years. Instead of using the term, "Terminator," it is now called the "Technology Protection System" (TPS). It implies a situation that is similar to an intellectual property claim which protects a novel invention. TPS involves the genetic modification and the subsequent spraying of a crop by a chemical so that the resulting seeds, which, though they appear to be normal, are not viable. Understandably, there has been much concern in developing countries, especially in Asia and Africa, with respect to the burden that farmers may have to bear in purchasing fresh seed for every crop. There have also been some incorrect statements which appeared in the press. Certain facts may be of interest.

The scientist who conducted the research which produced the "Terminator" gene technology is a biologist, Dr. Melvin Oliver, who has been involved in crop breeding and genetics for many years. In a recent conversation, Dr. Oliver stated that he has no personal economic motive in inventing the terminator methodology, which happened almost accidentally while engaged in related projects. A large seed company from Mississippi, called Delta & Pineland, filed a patent claim for the terminator technology because they provided funds for its research. Approval is pending. In 1999, Monsanto made a bid to take over Delta & Pineland, but so far it has not become a reality. Indeed, Monsanto has nothing to do with the invention of the terminator technology. One anticipated consequence of this new technology in the U.S. is to encourage farmers to use genetically modified and improved seeds for new crops. However, the possibility exists in the future to exploit the terminator technology to benefit farmers, where a voluntary choice can be exercised by an individual farmer whether or not to spray the plants with the antibiotic tetracycline since until doing so, the terminator is ineffective. A gene which acts as the "on-off" switch for the whole terminator system makes proteins that block production of the repressor proteins and have an affinity for a specific chemical, the antibiotic tetracycline. The spray induces the proteins to bind to the tetracycline instead of the enzyme production sites. When these seeds are planted, the DNA-cutting enzymes are made and the gene that makes the sterilizing toxins is primed to act when the plant matures. Some similar technology is considered necessary by certain experts to increase food production exponentially as the world's population continues to increase. Perhaps, Indian scientists will design their own unique brand of terminator technology which will suit the circumstances of the Indian farmer. The cost of improved seeds can be subsidized by the Indian Government or an international organization such as a branch of the United Nations. Some other applications of terminator technology are mentioned below.

## Controlling gene function

The entire biotechnology process, including sterility processes, has been under review by several companies and research teams, quite independently of Dr. Oliver's work. A new focus centered upon the use of chemical switches to turn specific genes in plants on or off at a critical time. Seed sterility is not the primary goal for much of this research effort. Controlling gene function directed at specific targets, e.g. enhancing oil content or nutrient value, is of greater interest, although it would, in turn, result in greater dependency of farmers on chemical companies. Others see the possibility of a new technology when a plant generates defenses against insects or viruses, by spraying an innocuous chemical that would then alert the crop to activate its normal defenses when threatened by an invasion or infection. It is hoped that this method will reduce the use of harsh pesticides normally used during the year. Similarly, instead of developing a new crop by means of genetic engineering to produce a toxin against insects, the new technology could create plants which would make the toxin only when threatened with serious damage. Limited exposure would make it more difficult for insects to evolve resistance.

The potential dangers of this new technology have been frequently emphasized in the press. Hope Shand of the Rural Advancement Foundation International (RAFI), a group based in Canada, wrote: "It's terribly dangerous. Half the world's farmers are poor and can't afford to buy seed every growing season. Yet they grow 15 to 20% of the world's food." But so far, the technology remains only speculative; its future development and practical application are yet to be decided.

Dr. Oliver has stated that the induction of sterility is not complete, but may reach about 70 to 80% in cotton. Another fact is that although much success has been achieved so far in cotton, experiments with tobacco have indicated a more difficult situation. Plans are afoot to experiment with all major crops eventually. It has been suggested that in such regions as Sub-Saharan Africa, recombinant DNA biotechnology and terminator technology are essential for the future survival of the populations. New technologies often bring forth adverse reactions, especially if there are implications of political and economic exploitation.

# ETHICS

# Ethics and Genetic Engineering

Much has been written about the ethical and moral implications of the use of genetic engineering technology in medicine and agriculture. It has been applied in developing therapeutic hormones such as insulin and methods for human gene therapy for treating various types of cancers and so on. However, progress has been slow and a major breakthrough has yet to be made in human medicine. Perhaps the most important outcome so far has been the mapping of the human genome.

It is a curious fact that the major opposition to the use of genetic engineering is in connection with agriculture and food production while medical applications of genetic engineering are generally regarded as more acceptable. This attitude may be a product of the trust traditionally placed in a physician and the belief that medical developments are generally in response to a perceived "need" that must be addressed immediately. No such sense of urgency is associated with the application of genetic engineering to agriculture. Many individuals and groups (some quasi-political) have argued that the improvement of crops and adequate food production can be attained without any recourse to the application of genetic engineering or genetic modification of any kind.

With these preliminary comments, one can assess the moral and ethical concerns in this regard in the following categories.

## (a) Ethical and moral framework

The moral basis for the ethical attitudes of the western world towards nature was strongly influenced, if not completely dominated, by the

Aristotlean approach, which implies that animals and plants are created solely for the use of human beings. That homocentric approach has not only led to the unlimited exploitation of animals and plants, causing many species to become extinct, but also to extreme cruelty in the treatment of various animal species. Of course, it has also led to the domestication of several species of animals and to the hybridization and selection of major crops.

The idea that humans are the ultimate masters of the world also led to the denigration of animals and plants as lacking rational thought or self-awareness and as devoid of any intrinsic value of their own. This view was perpetuated over two millenia by various philosophers and intellectuals as well as religious leaders. The rise of technology only intensified the belief that "nature" is here solely for the purpose of serving humanity, and new technologies, such as biotechnology, increased the control of nature by humans. Not only we are increasingly able to control "nature" and will perhaps do so absolutely in the near future, we are able to use the new tools of biotechnology to alter or manipulate the genome at a fundamental level that had not been possible in the past.

Underlying these developments, there is a faintly growing realization and a nagging suspicion by a few that the latest research indicates that animals at least are much closer to humans, both in their genetic and psychological make-up, than anyone had dared to admit until recently.

Thus, from a purely moral and ethical point of view, it is becoming clear that man has unjustifiably not respected the intrinsic rights of other living species on the planet. Issues of survival or matters of purely commercial interest seem to have dominated our approach to the understanding and treatment of "nature." This attitude has, in turn, led to various natural calamities, such as erosion, floods, depletion of biodiversity and extinction of rare species of animals and plants.

From this background, some have extended the argument that genetic engineering is a radically new form of intervention that poses a more serious challenge to our moral and ethical framework. Many are concerned that it is transforming "nature" in a direction that is irreversible and contrary to the inherent right of existence of all species on this planet. Others are troubled by transgenic technology, which make it possible to introduce genes across species, as a violation of God's creation. Yet others are concerned that a new moral dilemma is being created by genetic engineering because it could lead to a widening of the gap between rich and poor nations.

## (b) Eco-centric concerns

The relations between humans and other species and their relationship to their environments occupy a central place in an eco-centric universe. The well-being of the human species is closely entwined with that of other species and the rest of the universe. All living beings on this planet share a common evolutionary origin that is based on the properties of DNA. Any change, particularly an irreversible change, is viewed as being disruptive to this delicate balance of life, nature and environment. The impact of genetically modified plants and animals on the rest of the world is viewed with much apprehension. The ecological framework is generally perceived to be so delicate that any imbalance, such as genetic contamination through transgenic intervention, is viewed with much concern because such genetic alterations are considered to be irreversible. The impact of such intervention is believed to be disastrous for humans and other species as well as the entire environment including the ecosystems and the biosphere. Ecological ethic implies that the earth is a single ethical system, the ethical norm being the well-being of the comprehensive community, not just of human society.

## (c) Human health concerns

Health concerns are centered around the direct toxic effects of transgenic products and the indirect effects of a genetically contaminated world where the long-existing boundaries between the species disappear. Much concern is centered around the fact that recombinant DNA technology is considered imprecise. Gene targeting, introduction of the gene, and the control of gene expression in a new host are not based on mechanisms that are precisely known. There is a potential for carriers such as viruses and bacteria as well as segments of DNA to escape from target cells and infect nontarget organs and tissues, leading to cancer. Recombination between plant-produced viral genes and homologous genes of introduced viruses may produce new virulent strains which could infect a wider range of hosts.

Some ethicists have argued that genetic engineering will directly inflict much suffering on humans and animals because of the occurrence of various abnormalities in offspring. Similar defects could be caused by the teratogenic effect of transgenic foods as well. It should be emphasized, however, that while these concerns seem plausible, specific evidence in their support is lacking.

It has been suggested that animal models for studying AIDS infection using mice might prove dangerous because the AIDS virus carried by the experimental mice might combine with other viruses carried by the mice, creating a new and more virulent source of AIDS infection.

New proteins created by genetic engineering might result in allergic reactions in some consumers. Transgenic crops could introduce new allergens into foods, causing adverse reactions in unsuspecting consumers. One example is the case of soybeans that are genetically engineered to contain Brazil-nut proteins. It was found that those who are allergic to Brazil nuts showed allergic reactions to the GM soybeans because the allergenicity had also been transferred with the transferred gene. Furthermore, it is difficult to predict which proteins may induce allergic reactions in different individuals. Our experience with transgenic proteins has been quite limited so far.

Another source of concern is the presence of antibiotic resistance genes in foods. Such genes have been routinely used as markers in genetic engineering. Eating GM foods which contain these genes may reduce the effectiveness of antibiotics. It is further possible that the antibiotic resistance genes could be transferred to human and animal pathogens, rendering them resistant to antibiotics.

Other ethical and bio-medical concerns include the possible role of genetic engineering in increasing the levels of toxic substances. One source of this problem is the disruption of the gene complex that happens when an isolated gene is transferred. Associated genic interactions and epistatic interactions as well as various switch mechanisms are dislocated by genetic engineering of one or a few genes targeted for a specific function. New situations of increased toxicity may arise under the changed circumstances.

Genetic engineering may remove the natural protection of the plant in some cases. For instance, if genetic engineering is used to produce coffee plants with decaffeinated coffee beans by removing or switching off genes that produce caffeine, the plant could lose protection against fungal infection. This will, in turn, result in the production of harmful fungal toxins. Once again, it must be emphasized that transferring individual genes across species disrupts the natural order that is controlled by a holistic gene complex.

### (d) Biodiversity concerns

It is feared that the financial gains that are expected from genetically engineered crops may lead to the exclusive cultivation of such crops on a large scale to the exclusion of all other crops. It may thus lead to biodiversity loss, diminishing the available gene pool for breeding purposes in the future. Furthermore, pure lines of genetically homogeneous crops will be readily susceptible to a single infection by a bacterial, viral or fungal strain, destroying an entire crop instantly. Traditional crops with a genetically heterogeneous constitution do not suffer a similar fate. From an ethical and moral point of view, any technology that results in loss of life and loss of biodiversity is unacceptable.

### (e) Economic concerns

There are profound ethical and moral implications of economic domination and exploitation that are associated with genetically engineered crops and their intellectual property protection which is largely practiced in rich countries. This is particularly so in the case of the germplasm of both crops and their wild relatives as much of that biodiversity resides in the developing world.

As in the case of the terminator gene, some technologies may be seen as exploitative. The high cost of new technologies make them unattainable for many poorer countries. Regulatory agencies that are essential for monitoring the effects of genetic engineering in medicine and agriculture do not exist in many developing countries. Any damage caused by the new technology would be disastrous to debt-ridden developing nations which are barely able to feed their populations.

## (f) Impact on agriculture in developing countries

As already mentioned, many developing countries lack the appropriate regulatory mechanisms and resources to evaluate the risks involved in importing GM seeds or develop similar technologies in their own countries. On the other hand, they are the rich repositories of much of world's biodiversity which must be fully utilized for the benefit of their own people. It should not be exploited by developed countries which have the advantage of technological know-how and financial resources. There is also some concern that genetically engineered crops might displace the traditional crops of the developing countries, disrupting the lives of millions of farmers and others. These concerns are addressed in other chapters.

## (g) Regulatory concerns

Competent regulation to ensure safety is absolutely essential. Testing is crucial, yet in some countries genetically engineered foods do not require approval before marketing them to the public. Neither public notification nor labelling to inform consumers is required. Possible risk notification is also not required on the packaging. In the U.S. companies are allowed much latitude to determine if the risk level exceeds acceptable danger. Public protests in the U.S. against GM crops and foods have been relatively mild in comparison with those in Europe because the regulatory agencies have been weak in safeguarding the public interest. Consequently, the public was not made aware of the seriousness of the problem.

This is shocking in view of the fact that the area where GM crops are found in the U.S. is several times greater than that in any European country. Multinational companies such as Monsanto in the U.S. have been allowed to introduce GM crops on a vast scale without conducting any screening tests to assess the risks involved. Some critics have wondered about the revolving door between the U.S. Government and the biotechnology industry. Others have commented that the lobbying power of the industry is too great to be overcome by the regulatory agencies.

## (h) Political concerns

The political ramifications of the pros and cons of genetically modified technology are quite significant. It appears to have become yet another tool that is helping the rich nations of the world control and dominate the economies of poorer countries. The discoveries and inventions of new biotechnology with the accompanying intellectual property regime are perceived to be a serious threat to the agriculture of many developing countries. Within the rich countries themselves, GM technology is opposed by those involved in traditional farming and pro-environment groups such as the Green Party. Others have engaged in raising false alarms or panic over claims which are not supported by evidence. The fact of the matter is that neither side in this argument is backed by specific evidence. There is much speculation about the safety or risk involved in using GM technology. These issues are discussed in detail in other chapters.

## (j) Industrial monopoly concerns

Delta & Pineland is one of the largest seed companies in the world. The terminator seed technology was developed by Delta & Pineland in collaboration with the U.S. Department of Agriculture to produce a crop with inviable seeds so that no new generation can be cultivated. It was much criticized and attacked all over the world, but especially in developing countries where, quite erroneously, Monsanto (which was trying to acquire Delta & Pineland) was blamed for the invention. But the fear of controlling a new technology in the hands of a few remains valid. IPR regimes could lead to monopolies from new technologies, making them expensive and unaffordable to many.

## (k) Religious concerns

Transgenic technology where genes are transferred across species and kingdoms pose ethical and moral dilemmas. Those who are opposed to consuming the meat of certain animals (e.g. pork) may object to the incorporation of genes from those species into their diet. Adherents of certain religions do not eat pork. Many Hindus do not eat meat or fish of any kind and others do not eat beef although they may consume poultry or some other kind of meat. Vegetarians may object to the introduction of a gene from salmon into the potato or a gene from another animal species into their vegetarian diet.

Many religions believe in the sanctity of life as a unique gift of God. They object to its alteration by means of genetic engineering and its patenting as a commercial entity.

# The Third Culture, Ethics,
# the Missing Piece in the Puzzle

Kevin Kelly stated that "science has always been a bit outside of society's inner circle," but did not go into the possible reasons to account for that situation (*Science*, February 13, 2000, p. 992). Furthermore, it is questionable whether technology, the so-called "nerd" culture, represents a truly distinct entity that is independent of science. It does not appear to involve the same degree of distinction which exists between the humanities on the one hand, and the sciences on the other, which C. P. Snow[1] had first recognized. Early science writers such as H. G. Wells made no distinction between science and technology. Julian Huxley[2] saw evolutionary humanism as a connecting line which could bridge the culture gap discussed by Snow. Huxley had suggested that the time was ripe to form a comprehensive, coherent and unitary pattern of ideas in this regard, and that education of the future should be centered around the idea and the facts of evolution, a core-biology connecting physical science with psychology, evolution linking humankind with the rest of life, and history as a psycho-social evolution.

Ever since the dawn of science in its present form, scientific thinkers have innocently assumed that a society's value conflicts are largely the result of ignorance which can only be corrected by an extension of the scientific method and its value system to the rest of the society. Jacques Monod,[3] for instance, has suggested that the "ethics of knowledge" from a scientific viewpoint implies a responsibility to lay the foundations for a value system which is wholly compatible with science itself. At least one political scientist, Yaron Ezrahi,[4] has pointed out that this approach implies a readiness to assume the possibility of a homogeneous value environment for an unproblematic application of science to social problems. The belief that the criteria of scientific and technical rationality could be universally applicable outside the context of science is already shown to be impractical time and again through practical experience. Yet, generations of scientists and writers continue to return to this hope of the universal application of a scientific rationality. But, is it true that science and technology are readily accepted universally? It is hard (if not impossible) to bridge the gulf between the pluralistic competitive value environment of democratic public policy-making and the relatively homogeneous consensual value systems of professional scientific societies. It would be naïve not to recognize that any attempts to extend the scientific standards of reliability, validity and rationality to the larger context of public affairs are only fraught with frustrations and failures. Closer to home, it is even more clear that scientists themselves differ widely on the desirability of the applications of science and technology to society.

There are other reasons which explain why "science has always been a bit outside of society's inner circle." Quite inadvertently, science and technology have introduced new forms of threat to our future security. The possibility of biological and chemical warfare, among other threats, gives us pause in this context! Kelly's enthusiasm for technology includes a passing reference to its disastrous consequences, and he recognizes that technology is not going to solve our social ills. There is yet another aspect which was not mentioned by him. Joshua Lederberg,[5] among others, has drawn attention to "a kind of imperialism of the present over the future, a closing of options that our children should have available to them... a legacy of technology, about whose merits they had no voice in deciding. So our prior ethical task is to what is owing in intergenerational responsibility."

Other ethical problems arise when technology goes global. A different kind of technological imperialism involves the transfer of technology to developing nations who cannot afford the resources and skills needed to protect their citizens from the adverse consequences of technology. Far from curing social ills, technology has created a great number of environmental, biological and other hazards which are being passed on to future generations of humans and other species. For these and other reasons, if technology has indeed seized control of the culture, as Kelly states, then we are in far greater trouble than is widely recognized. It is true that technology has opened up new opportunities, but it has narrowed our options as well, both for ourselves and our descendants.

# Haldane's Dilemma and Its Relevance Today[*]

## Introduction

The scientific and intellectual eminence of John Burdon Sanderson Haldane is such that we are indeed fortunate to be holding this symposium under his name. The very mention of his name evokes scientific excellence and its benefits to humankind. Indeed, by honoring Haldane, we are honoring ourselves.

It has been said that it is only by standing on the shoulders of giants that we can see far. J. B. S. Haldane was one of those giants. Even some thirty-seven years after his death, Haldane's direct impact can be readily seen from the long list of citations to his publications in the Science Citation Index and similar compilations today. Both directly and indirectly, Haldane's world-wide influence has been quite significant even from the time of his first scientific paper — in collaboration with his father, John Scott Haldane, in respiratory physiology, which was published in 1912. To this day, he remains one of the most quoted scientists of the twentieth century. He left so much unfinished work when he died in 1964 that his manuscripts continued to be discovered and published posthumously. The last publication was a book on science fiction called "A Man with Two Memories."

---

[*]Delivered at the M. S. Swaminathan Research Foundation, Chennai, India, January 1997.

# Haldane in India

J. B. S. (or Jack) Haldane was born in Oxford, England, on November 5, 1892 and died on December 1, 1964, in Bhubaneswar, India. He received his formal education at Eton and Oxford, studying mathematics and classics, but received no academic qualification in science. However, his early scientific education was provided at home by his father, distinguished Oxford physiologist John Scott Haldane. The younger Haldane successfully (and successively) pursued research in physiology, genetics, biochemistry, statistics and biometry at several English universities including Oxford, Cambridge and London. In addition to his brilliant scientific contributions, Haldane became famous for his outstanding popularization of science in the lay press. Furthermore, he took part in politics, especially during the 1930s and 40s, writing and speaking extensively on Marxist philosophy. It was during that period, from 1957 until his death in 1964, that I came to know him intimately as a pupil of his and as a colleague in Calcutta and Bhubaneswar. He adopted an Indian lifestyle — wearing Indian-style clothes, eating only vegetarian food, studying Hindu classics and mythology, etc. All of these distinguishing qualities, intellectual, political, philosophical and personal, combined with his highly interesting family background, made Haldane a fascinating character for biographies.

I have previously described the biographical details of Haldane's life and work in numerous books and papers.[1-9] These may be consulted for further details. Haldane himself left a highly informative autobiographical sketch which mentions his life in India and his association with myself and others.[10]

In India, Haldane was associated with three different institutions: the Indian Statistical Institute in Calcutta (1957–1961), the Genetics and Biometry Unit of the Council of Scientific & Industrial Research in Calcutta (1961–1962), which he tried to establish unsuccessfully, and the Genetics and Biometry Laboratory of the Government of Orissa in Bhubaneswar (1962–1964). The topics covered by Haldane and his associates included human genetics, plant breeding, biometry, animal behavior and theoretical or mathematical population genetics. In addition to these activities, he lectured widely and contributed popular articles to the lay press.

## Intellectual hybridization

A biographical analysis of Haldane's intellectual growth and achievements indicates that, because of his extensive knowledge of several branches of science, he was able to create new concepts by synthesizing ideas and concepts from several existing disciplines. I have called this process "intellectual hybridization." Examples include the foundations of "population genetics" and "biochemical genetics," particularly, the gene-enzyme concept, "immunogenetics," "gene mapping," "molecular biology," "behavior genetics," etc. Haldane played a pioneering role in all of these fields. In his classic, *The Causes of Evolution*, Haldane was the first to analyze the evolutionary consequences of altruistic genes in populations (e.g. honey bees) which later led to the founding of "sociobiology" by W. D. Hamilton and E. O. Wilson. It was also in that same book that Haldane estimated the first human mutation rate (for haemophilia). Haldane's contributions to the foundation of human and medical genetics were so important that he had no parallel. These are but a few examples which I have selected from Haldane's numerous scientific contributions.

## Science and ethics: Haldane's dilemma

Although Haldane has made several profound contributions to physiology, biochemistry, genetics and other sciences, he may well be remembered by posterity for his advocacy of an adequate ethical framework that would be able to incorporate new developments in science and technology, especially genetic engineering. In his book, *Daedalus, or Science and the Future*, Haldane[12] was the first to raise a number of ethical dilemmas which are the focus of attention in biotechnology today. His original work has been recently reprinted in my book: *Haldane's Daedalus Revisited*.

In 1923, Haldane had already foreseen the kind of disasters — both in war and peace — which resulted as consequences of technology in this century.

## Genetic engineering

Haldane's predictions in *Daedalus* include the widespread use of *in vitro* fertilization in human societies, resulting in the production of a large proportion of test tube babies each year, artificial transmutation of genetic material, and practices of eugenic selection to produce individuals with exceptional skills in music, art, sports, sciences and other disciplines.

Haldane's scientific predictions go hand-in-hand with dire predictions of social and ethical disaster if their applications are not made with a great deal of caution. Advances in reproductive biology and genetic engineering today have already seen certain misapplications of adverse impact. Fetal selection and infanticide have been practiced in certain communities. Products of genetic engineering such as high milk-yielding cattle and various crops have not met with social acceptance as had been anticipated. There is also continuing fear that genetic engineering might be used for the purpose of developing weapons of biological warfare. For this reason, some developing countries have expressed reluctance in participating in collaborative research which involves research on DNA polymorphisms.

Genetic engineering and other aspects of modern biotechnology have raised a host of new issues in trade and international relations. There is no general agreement about the nature of patenting of biological materials and processes. Most countries have had no domestic legislation regarding intellectual property rights until recently, but, under the GATT agreement, this situation is expected to be remedied soon. There continues to be profound differences among individuals and nations regarding the patentability of the human genome. Patenting of naturally-occurring DNA sequences should not be allowed (as some in the U.S. have suggested). Specific processes and products may be patentable under certain conditions.

Shared credit for intellectual property rights presents an especially difficult problem. Among other problems are the rights of the indigenous people such as the farmers who pioneered agricultural practices a long time ago, utilization of raw materials, and parallel development of various practices and products. Technological progress proceeds at a different pace in each country. This depends on several factors: financial resources, level of technology and education, government policies and domestic business, and intellectual climate. There is a basic question of ethical judgment; whether a few technologically advanced countries can appropriate the world's resources and claim intellectual property rights which would be detrimental to the rest of the world. This is especially relevant to the large number of developing nations which contain many of the raw materials and are populated by a majority of the world's population.

## Problems with new technologies

In spite of the fact that new technologies usually create new benefits, a note of caution is warranted. Important lessons can be learned from the experience of technologically advanced nations. At least four instances can be cited: (a) If new technologies are introduced too quickly, increased unemployment may be the result. Gradual transfer or creation of technology should go hand-in-hand with extensive training and literacy programs; (b) Technology transfer (or sharing) by its very nature, creates new social values and standards which often conflict with traditional beliefs and traditions. It may pose a threat to indigenous economies and social systems, thus turning the whole matter into political controversy; (c) Biotechnology involving genetic engineering and reproductive biology may be misused on the one hand (e.g. aborting a fetus for trivial reasons) or may lead to new social problems. Potential carriers of genetic diseases may face difficulties in obtaining education, employment, and medical or life insurance, and; (d) New technologies may occasionally pose risks for public (or individual) health by means of biological or chemical contamination of the environment or by increasing stress levels. A special problem may be created by the long-term consequences of human gene therapy. The long-term carcinogenic possibility of recombinant DNA, when introduced into the cells of patients, is not yet known.

The genetic revolution predicted by Haldane has arrived. It came twenty years later than he thought. We should consider ways of resolving Haldane's dilemma. Instead of stumbling into problems created by technology, we should have a deliberate coordinated program to anticipate potential problems of impending technology and find ways of resolving them before they become active crises. This is one of the lessons we have learned from the experience of the U.S., Europe, and other industrialized parts of the world.

## Anticipating the consequences of technology

Educational curricula should include courses on the impact of technology and its potential consequences. But this is not the case in many countries. The coordinated and planned introduction of technologies should be preceded by simulated computer analyses of the potential impact of new biological, chemical or other products on the health of future generations (such as the projected contamination levels of air and water), public education of various issues by using television and other media, projected costs and available resources for coping with the consequences, anticipation of adequate employment and training strategies, and a choice of alternative technologies based on cost/benefit ratio, etc. These decisions should be based on long-term considerations and not on short-term benefits.

We honor Haldane by remembering this duality which he emphasized long ago. On the one hand, we cannot go very far without the benefits of science and technology in today's world. On the other hand, we should find ways of anticipating and coping with the economic, biological and other consequences of new technologies. In both situations, we will surely benefit by following Haldane's admonition.

# Cloning Again!

Cloning is in the news again. This time it involves the cloning of mice. Scientists at the University of Hawaii reported that they have produced more than fifty mouse clones. This was mainly achieved by Dr. Ryuzo Yanagimachi and his postdoctoral student Dr. Teruhiko Wakayama. This is a very important development because it confirms the earlier work of Ian Wilmut who produced Dolly, the sheep clone, in Scotland. Cloning has become almost a routine process now. What is next? Human cloning? It is of interest to speculate on the possibility of cloning humans and its practical applications.

The ability to create copies of humans has long been the substance of science fiction. The technique of nuclear transfer allows the reconstruction of any embryo by the transfer of genetic material from a single donor cell to an unfertilized egg from which the genetic material has been removed. Some concerns about cloning have been expressed as follows: (a) humans will be treated like other experimental animals for laboratory research, (b) a market in fetal tissue will develop, transforming abortion into a medical parts industry (thus denying human life of any sanctity), (c) a rush to "immortality" will lead to a black market in spare body parts, draining much needed resources for legitimate medical treatments, (d) cross-species experimentation (which is now forbidden by some religions) will be encouraged by the cloning methods and will be supported by commercial markets, and finally, (c) these developments may lead to increased interference in personal lives by governments, resulting in restrictive legislation concerning our future genetic heritage.

Under these conditions, the sanctity of life will lose all its meaning. Like automobiles and real estate, life and living parts will become just another commodity. They will be subjected to stock trading and other commercial practices like any other product. Federal regulations will have to be modified to monitor the growth of this new biotech industry.

The life of an individual, as we normally conceive it, is defined by certain boundaries and parameters. So far, our genetic lineage has been strictly defined and passed on to the next generation. However, under the new methods of cloning biology, these rules are broken. There is a flexibility of genetic transfer in several directions between species, between generations, and between individuals. It will create new and exciting opportunities which are unimaginable at present. The new cloning biology will help us to eradicate diseases and create new generations of populations which are genetically modified to possess exceptionally superior qualities in all fields of human endeavor.

# Stem Cell Research: Which Way to Go?

Powerful lobbies in Washington and the nation have been debating the pros and cons of allowing stem cell research to go forward. For the U.S. government, a positive response involves not only abandoning any idea of banning such research, as some have advocated, but also making generous funds available to support the research on stem cells in various research and teaching institutions across the country. The debate is being carried on at several levels — theological, ethical, political and scientific. For a scientist, the decision is fairly simple. If the research is scientifically sound and is likely to yield beneficial results, then one must go forward with the project. However, as we shall see, because of the use of fetal cells in research, other factors enter the picture.

What is stem cell research? Stem cells are the earliest cells found in the human embryo before they become differentiated to develop into specialized nerve cells, muscle cells or any one of a large number of potential organ tissues. Research on stem cells can potentially lead to a better understanding of how various human organs develop, how errors are made in nature resulting in birth defects or other defects or diseases that appear in later life, or how to transplant cells to cure such diseases as Alzheimer's. Obviously, stem cell research is of vital importance as it can open doors to cure not only Alzheimer's, but also diabetes, Parkinson's, various cancers and many other diseases which cause much suffering and death to millions of people today.

The Clinton administration issued guidelines allowing federal funding for research on fetal stem cells that were taken from frozen human embryos derived from *in vitro* fertilization, which would normally be discarded after the treatment of infertile couples. However, anti-abortion forces argue that research should be restricted to adult stem cells, which can be harvested without destroying embryonic life. On the other hand, some conservatives in the Republican Party favor research using human embryos. They include Senator Orrin Hatch of Utah, the ranking Republican on the Senate Judiciary Committee and Senator Connie Mack, a Roman Catholic, of Florida.

Several highly placed individuals in the Bush administration and The U.S. Congress are said to be in favor of stem cell research, mainly because of the occurrence of some dreadful diseases in their families. Some Republicans are said to be in favor because of the long suffering of former President Ronald Reagan from Alzheimer's. Bush has finally decided to allow only limited research with stem cells that can utilize cell lines that are already stored in U.S. laboratories but involves no new cell lines. No doubt research will still go on at private institutions and in other countries in Europe and Japan where resources for less restricted research are available.

# Race and Human Dignity

As a human being and a geneticist, I find it shocking that some authors like Dinesh D'souza are arguing, for political reasons, that such experiences as slavery and other indignities are only a figment of our imagination.

This is clearly nonsense and we must not let it pass without a counter argument. Such people tend to blame the victim, rather than the perpetrator of the crime. Racist propaganda takes many forms. Under the guise of something else, it tries to acquire a façade of "respectability" which it does not deserve.

Some of these politicians may even get elected to high political offices, thus finding many opportunities to spread their racist propaganda. If we follow the racist propaganda, then we have to accept the old colonial argument that the people of India and other countries deserve to be ruled by alien powers because they were of a genetically inferior race.

The colonial rulers saw their atrocities only in terms of benign colonizers. Propaganda back home in Britain, France and other colonial countries told their people how they were helping to civilize some primitive populations all over the world. Whole generations for centuries grew up believing this nonsense and continued to support their governments' colonial policies.

Now, you might ask what all of this has to do with us today. The answer is quite simple. Such human indignities as slavery and colonial domination clearly leave their mark on the psyche.

The resulting damage lasts a long time. Even though we try to recover from this past inequity and continue to build nations with a renewed vigor, certain fundamental attitudes and behavior patterns have become entrenched in our society.

These considerations bring us to pay more attention to such revisionist historians as Dinesh D'souza, because their misguided opinions continue to influence readers. As I said before, racism takes many forms.

In today's America, racist attitudes are reflected in the budgetary priorities which are proposed each year by the U.S. Congress. They are also reflected in foreign policy. Friendly lobby groups are helpful because they can help to present a more balanced perspective. Both domestic and foreign policies of a country may often reflect subtle racist attitudes.

The word "race" may not be used, but it is reflected through economic policies, budgetary priorities and treaties with other countries.

What can we do as individuals? We should certainly present counter arguments against "racism," whether it is overt or subtle, or at the individual level or national level.

We must also examine our own individual behavior patterns and attitudes to make sure that we are not becoming "players" in a racist game without being aware of it. We must also educate ourselves and others on the role of genetic and environmental or social factors in shaping our destiny. Through continued education and vigilance, we must counteract the evil consequences of racist propaganda. If those in positions of power like legislators and heads of governments are educated in human biology and genetics, then racist attitudes of the past and present will become a distant memory.

# DEVELOPING COUNTRIES

# The Politics of Biotechnology

The terminator gene, neem tree products, basmati rice, the medicinal properties of turmeric and other Indian plant species; what do they all have in common? They have become a part of recent international controversies in the context of intellectual property rights and the use of biodiversity.

In recent years, genes have been given human attributes with such names as the "selfish gene," "the traitor gene," "the cheating gene," "the altruistic gene," etc. I am not sure how valid such nomenclature is because genes have no motivation in the human sense. Many fear that the "terminator gene" will start a new cycle of exploitation.

World War II spelled doom for the old imperialist countries. In the following years, led by India, many poor countries of the world won their hard fought independence from colonial rule. But the same colonial powers, now led by the U.S., have been attempting to establish a stranglehold on the developing countries in the form of trade and tariffs. In this prolonged struggle for exploitation, intellectual property laws have become just another weapon. They are replacing old style colonialism. The present form of IPR laws have been entirely designed by western countries. They are well suited to the economic benefit of rich countries.

For instance, take the case of crop germplasm. Prof. Noam Chomsky wrote: "For thousands of years, people in the south have been developing crops. They don't own them. They don't get any rights from them... They have the rich gene pool and thousands of years of experience in creating hybrids and figuring out what herb works. Then western corporations go in and take it for nothing. We minimally modify it and sell it to them. We patent it. It is a scam designed to rob the poor and enrich the rich." (*World Orders, Old and New*, Columbia University Press, New York, 1994).

Developing countries are slowly awakening to the great need that they must design their own intellectual property laws that will guarantee their best interests in the international arena.

## North-South struggle

Biotechnology is the latest tool in the war that has been going on for several centuries. Globally speaking, it is a war between the rich countries of the north and the poor countries of the south. Countries of the south are rich in biodiversity and natural resources.

Economic supremacy is the goal and reward of this struggle. IPR laws are designed by the north to perpetuate the *status quo* of this unequal economic partnership. All of the resources and strengths of the countries of the north are directed toward this goal.

For instance, since World War II, the U.S. has steadfastly entered into conflicts (either directly or by proxy) with poor countries including Korea, Cuba, Chile, Iran, Iraq, the Dominican Republic, Vietnam, Cambodia and Yugoslavia. In some cases, as in Africa, they have entered into a partnership with Great Britain, utilizing its resources in the former colonial countries. Whatever political ideology or excuse may have been used in each instance, the underlying goal is always the same: the struggle for the control of natural resources and economic exploitation.

## The recent truth

The Commission Report in Guatemala has revealed that the CIA was directly involved in a program to eliminate 2,000 communist insurgents, but actually slaughtered more than half a million civilians, including some nuns and whole villages of Mayan Indians. One wonders as to the real basis for these conflicts! Even where ideology is mentioned, it is usually transparent that economic exploitation by the rich countries of the north is always implicated.

So we seem to have come a full circle. Colonial empires were destroyed, but they are now replaced by economic exploitation in the name of IPR and new technologies. Political ideology, colonial administration, trade and tariffs, IPR, or whatever excuse one may think of, the end result is the exploitation of the developing countries of the south by the rich countries of the north. It is feared that genetic engineering is the latest means that is being employed in this struggle.

India will eventually develop its own particular brand of intellectual property rights which will suit its best interests. This will be especially necessary as domestic technologies are increasingly developed, utilizing local resources and skills. However, the problem of recognizing the intellectual property of traditional farmers and the medicinal knowledge of indigenous tribes still remains.

As Professor M. S. Swaminathan puts it: "It is thus appropriate to conserve not only agrobic-diversity, but also the traditions which led to the enrichment of our knowledge on valuable genetic material. Tribal taxonomy differs from modern biosystematics in the sense that in most cases, the tribal nomenclature has its roots in the end use of the material. It is thus value added biosystematics." (From: *Agrobiodiversity and Farmers' Rights*, Konark Publishers, Delhi, 1996.)

# The Economics of Biodiversity

A discussion of the economics of biodiversity must start with the following facts:

(a) Biodiversity is of commercial and economic value,

(b) Economic factors and population growth are the main driving forces behind the extinction or near-extinction of many species of plants and animals,

(c) Long-term conservation of biodiversity yields maximum economic benefits,

(d) Shortsighted exploitation of biological resources not only leads to the destruction of biodiversity but does not result in higher economic benefits,

(e) Inadequate and inappropriate planning of biodiversity utilization is due to shortsighted policies and a lack of appreciation of the value of biodiversity utilization, and

(f) While developing countries possess much of world's biodiversity it is the developed countries, with their technical knowledge and resources, which benefit most from the biological wealth of the planet.

The politics of biodiversity utilization are closely linked to the social and commercial developments which evolved during the years following World War II. The emphasis on the resource intensive and labor-intensive strategies of developing countries to promote their exports has proved destructive to both the environments and the economies of those countries. Developing countries view global environmental issues as the result of extended industrial outputs over centuries. For instance, greenhouse gas emissions arise from the use of energy. It is estimated that seventy percent of the world's $CO_2$ emissions and most of the CFCs are emitted by industrialized countries which contain only about twenty-three percent of the world's human population. On the other hand, developing countries, with seventy-seven percent of world's population, account for only thirty percent of the world's emissions. That is to say, in per capita terms, the emission of carbon is ten times higher in industrial countries than in developing countries. The U.S. emits about twenty-five percent of all carbon emissions while consuming about twenty-five percent of all petroleum produced, even though its population is only about five percent of the world's population. If this pattern were to be followed by the two countries with the largest populations, namely China and India, it could potentially result in a great economic and environmental global disaster.

A historical review of the post-World War II years indicates the manner in which the present state of affairs has emerged. The war was won by the allies, but the leading European countries and Japan suffered a serious economic setback. Their economies, infrastructure and industrial base were almost totally destroyed. The U.S. emerged as the preeminent economic power during those years, producing about forty percent of the world's products. (Today, its productivity has come down to twenty-five percent of the world's economy, measured in terms of the Gross National Product (GNP)). During the post-war years, four major international organizations were created: the United Nations (UN), the International Monetary Fund (IMF), the World Bank (WB), and the General Agreement on Tariffs and Trade (GATT).

In the emerging new world order, their agendas and programs reflected the leading economic power's vision of growth, such as "an extremely resource-intensive growth corresponding to a rapidly expanding frontier economy, with an enormous consumption of resources, and the domination of nature through rapid technological change." (Chichilnisky, 1998).[*] Two major economic theories of trade have

---

[*]*Sustainability: Dynamics and Uncertainty.* Eds. G. Chichilnisky *et al.* (Kluwer, Boston, 1998).

prevailed during this period. The first is the neoclassical theory of optimal economic growth (originating in the U.S.) which views a steady, long-run path of development in terms of exponential rates of population growth and correspondingly exponential increases in resource utilization. This theory implies an unlimited expansion in terms of the economy and its use of resources. Secondly, the theory of comparative advantages in international trade, which originated in Sweden, recommends that developing countries emphasize exports of resources and labor-intensive products and trade them for the capital- and technology-intensive products of developed countries. These two theories have had a decisive influence during the last fifty years in shaping the policies of the World Bank and the International Monetary Fund toward the developing countries. These agencies provided strong incentives to developing countries to export more and more resource-intensive products to qualify for loans and technical aid. Even more importantly, economists, civil servants and other intellectual leaders in the developing countries, who received their training in the U.S. and other western countries, have been imbued with the unwavering sense of promoting the agenda of the World Bank and the International Monetary Fund. Their primary approach underlies the belief that developing countries are only good enough for cheap labor, natural resources and cash crops. Much of the world's population lives in the developing world.

## Social and cultural values

Appreciation of nature, including all its habitats and other variations, was succintly stated by Virgil:

> It is well to be informed about the winds,
> About the variations of the sky,
> The native traits and habits of the place,
> What each locale permits and what denies.

In short, biodiversity is a part of nature, and nature connects us all in myriad ways — from the viewpoint of aesthetics, knowledge and commerce. Biodiversity is exemplified in our poetry, music, paintings and other art forms. By understanding nature, we understand ourselves.

We are accustomed to viewing it as a part of ourselves, or conversely, we are part of nature. Our own origins are inextricably linked to biodiversity.

As a consequence, there is a presumptuous sense of ownership on the part of all human beings with respect to what we call nature and all its contents. In many developing countries (with subsistence economies) biodiversity has been (and is) viewed as a collective property resulting in benefits that are shared by all. The concept of individual ownership and intellectual property in the context of biodiversity, although being debated actively in recent years, still remains a novel idea in many developing countries.

Our values, within the context of biodiversity, are further defined by our religious and moral beliefs. All mythologies contain multiple references to various animal and plant species, and in some religions many of these are believed to possess divine qualities. For instance, in Hinduism, various animal forms are worshipped as Gods and plants such as *Ocimum sanctum* (*Tulsi*), which are revered for their mystical and healing properties.

## Problems of valuation in subsistence economies

Valuation of ecosystem services presents several problems that are unique to subsistence economies (see Daily, 1997).[†] Traditional western approaches to valuation on the basis of indirect use values and non-use values are not applicable to societies with radically different social and economic structures not only because of methodological difficulties, but also due to the moral and ethical dilemmas they pose.

- Many societies with subsistence economies are very heterogeneous, with varying systems. They are neither fully monetized nor adequately informed about choices to exercise various options.
- Certain traditional benefits enjoyed by these societies, such as the use of herbal medicines by the local people or the extent of prevailing intra-family care in nursing the sick and dying, are not quantifiable in monetary terms.

---

[†]Daily, G. C. (Ed.) *Nature's Services* (Island Press, Washington D.C., 1997).

- Furthermore, a complex system of biotic interactions, involving timely cooperation between various animal and plant species, micro-organisms, soil conditions and climatic factors, is responsible for the productivity of ecosystem goods. Current valuation methods do not assign specific values to these processes or to such biological factors as pollination, seed dispersal and nutrient uptake, etc.
- Another problem is the valuation of cultural, religious and spiritual services rendered by an entire ecosystem or a specific species. There is no precedent in the methodology of valuation to include such services.
- There is also the obvious failure to take into account the large mass of traditional knowledge that has led to many important developments in agricultural practices, pharmaceuticals, crop and animal breeding, toxicology, horticulture, forestry, and the commercial production of various animal- and plant-derived products.
- Current valuation methodologies do not take into account the extreme heterogeneity found among various sections of human society with respect to their perception of the degree of importance attached to different aspects of life and living standards, for instance, those living in big cities as opposed to others in the rural areas.
- Subsistence level benefits are undervalued in situations when the destruction of an ecosystem leads to population displacement and great human misery (eco-refugees).

Any discussion of valuation involves the possibility of commercialization and the ensuing deleterious effects on biodiversity. Many forest products, such as rattan, are becoming scarce. Large numbers of workers were employed in the rattan trade, including over 150,000 in Indonesia alone, and an international trade value of almost US$7 billion. The tendu leaves which are used in India to wrap tobacco for making *bidis* (a cheap cigarette) generate an annual revenue of US$160 million in one state alone (Madhya Pradesh). The contribution of ecosystem goods and services to the GDP (Gross Domestic Product) is often undervalued in a subsistence economy. The annual rent from forests for both goods and services exceeded more than twenty-five percent of the GDP. It was pointed out further that ecosystem products (in particular, fuel-wood and fodder), which have been the foundation for subsistence economies, contribute approximately thirteen percent of the total value of forest goods and services. It was estimated that the annual value of non-timber forest products and services from dry deciduous forests in India amounted to about US$220–US$335 per hectare. These include soil conservation, nutrient cycling, tourism and recreation. The GDP estimates in the past had not included the value of these services. India is not alone in this respect. GDP estimates for several developing countries, including Brazil and Mexico, suffer from the same fate.

## Economic value of biodiversity

It is a strange paradox that much of the world's biodiversity is in developing countries whereas the theory and practice of economic evaluation has been developed and applied mostly in the developed world. However, it has been suggested that these practices may not be applicable universally due to the absence of "freely" functioning markets for inputs (e.g. labor, capital, raw materials) and outputs (e.g. agricultural produce) in developing countries (Pearce and Moran, 1994).[‡]

There are several different ways in which biodiversity may be regarded as a source of great value to human society:

(a) Biodiversity is a source of knowledge which can be exploited for biological prospecting and economic benefits (e.g. medicinal products).

(b) Biodiversity is the source of all food supply, which feeds and sustains all life on the planet.

(c) Biodiversity is a vital part of the life-support systems on earth which sustain human life. In addition to food production, it provides other life-sustaining elements such as the production of oxygen by green plants, the cleansing action of bacteria in various contexts and the nitrogen fixation in the soil, etc.

(d) Biodiversity is the source of many cultural and aesthetic values and symbols. For instance, the significance of elephants, cows and other

---

[‡]D. Pearce and D. Moran. *The Economic Value of Biodiversity* (Earthscan, London, 1994).

animals as well as certain plants (e.g. *Ocimum* spp. or *tulsi*) to Hinduism, the bald eagle for the U.S., and so on.

(e) Biodiversity is the source of all biological variation, which can be manipulated by genetic engineering to create new varieties and breeds of various plant and animal species tailored to specific needs.

(f) The DNA revolution has revealed the universality of the underlying genetic mechanism of all life, a finding of enormous scientific and philosophical importance, which has revolutionized our concept of life.

(g) An understanding of the mechanisms which underlie biodiversity and genetic variation is contributing valuable knowledge for the treatment of genetic diseases.

(h) A study of human biodiversity has increased our knowledge of the basis of human variation, which, in turn, has promoted equality and tolerance among human populations.

(i) Biodiversity, by contributing to the ecological diversity of the planet, is a major contributor to eco-tourism.

A discussion of biodiversity has helped to define and refine our concept of intellectual property rights and the great value of biodiversity to developing countries. Heal (1998)[§] singled out four major economic issues that are central to our understanding of the extent to which market forces can define the value of biodiversity. These are discussed below.

## (a) Property rights

Only property rights can empower an individual (or an organization) who is the true owner to sell goods and transfer them to others. Experts on law and economics have discussed this subject in detail. However, biological resources pose certain special problems because they are not owned in the conventional sense. Biodiversity in itself is not a tradable commodity even when the ownership of biological resources is established. Hence, it is hard to appropriate their market value. In

---

[§]G. Heal. *Valuing the Future: Economic Theory and Sustainability* (Columbia University Press, New York, 1998).

recent years, legal attention to biological issues has been expanding. International agreements have been either proposed or ratified in relation to endangered species, wetlands preservation, salmon fishing, the life of the antarctic, and other living resources that might migrate across international boundaries. Many laws aim to protect regenerative resources. The Convention on Biological Diversity (CBD) has been motivated by the concern for the portfolio value of genetic material. It has been defined as the long-term insurance for human health and welfare.

### (b) Public goods

Biodiversity is a public good, i.e. a property that is shared by all. Examples of other public goods are law and order and defense. Traditional wisdom tells us that markets work best for private goods. Sustaining genetic variation and its marketing requires a different type of strategy. It is a commodity which is being depleted steadily. At any given time, the total biodiversity remaining in the world is the end product of natural and artificial selection, the latter including numerous decisions, over thousands of years, impacting on the nature and variety of crops and fruit trees, etc. that needed to be grown, the type and area of land that could be utilized, and how much of the wildlife habitat must be saved and so on. It surely encompasses the intellectual properties of countless tribes, indigenous communities and early inhabitants whose private decisions contributed to the ultimate public good. Biodiversity is an example of public good to which economists have not paid sufficient attention.

### (c) Sustainable levels of biodiversity

Some experts have suggested that the level of biodiversity that is needed to be maintained can be determined by the utilization of "depletion quotas" and "time horizons" (Heal, 1998). Any targets that are agreed upon are ultimately determined by the estimated total social costs and benefits of such a decision. But there have been no systematic analyses of these issues. On the basis of the relationships between populations and minimum thresholds needed for survival, a categorization of endangered species can be attempted. However, what is even more important is the forecasting of future threshold levels of biodiversity that is absolutely necessary to sustain a population at different levels of density as well as at a certain level of qualitative standard. As the population approaches the threshold, acceptable reductions or "depletion quotas" may be agreed upon, as, for instance, in the case of certain fisheries, there are standard methods of determining acceptable catches each year. Such quotas are a tradable commodity in an open market and may be used in such activities as the management of game and fisheries, forest clearing, and land management as well as irrigation. Incentives have to be created to encourage lower depletion rates and there is a market value involved which can be traded on world markets.

## (d) Tradable commodity

Time-dependancy of all biodiversity estimates make it difficult to make accurate forecasts of future trends. It is almost impossible to utilize conventional economic criteria to justify any investment decisions regarding biodiversity. There are too many variables which interfere with the forecasting process. These include climate changes such as global warming, soil and water contamination, leakage of nuclear waste and other disasters for which humans are responsible. A proper economic valuation of biodiversity should be an integral part of all projected social and economic planning.

Biodiversity as a tradable commodity is unique in the sense that we cannot depend on the market to preserve it. One has to wait for the appropriate time when a correct valuation of biodiversity can be made. Otherwise, it is quite easy to undervalue biodiversity. The usual market approaches are quite useless to the process of evaluating the value of biodiversity or its forecasting. Some aspects of biodiversity *are* marketable. However, they are only a small part of the sum total value of biodiversity on earth. Intrinsic, aesthetic and cultural values of biodiversity are not protected by the market but by legislation such as the Endangered Species Act or the *Convention on the International Trade in Endangered Species* (*CITES*).

## Global valuation of biodiversity

Among a number of approaches to the assessment of biological value, the following criteria have been singled out by a committee of the U.S. National Academy of Sciences' Board on Biology (*Perspectives on Biodiversity: Valuing its role in an everchanging world* (National Academy Press, Washington, D.C., 1999).

(a) *Richness.* The number of species or habitats in a given area,
(b) *Endemism.* The narrowness of the distribution of the species in an area,
(c) *Rarity of species or habitats in a region.* A region with rare species or habitats is given higher value than a region with abundant ones,
(d) *Ecosystem services.* The importance of the natural habitat (even if only one species) with reference to its impact on the ecosystem, especially for various services provided to the human population, and
(e) *Protected status and representation.* The protection of the ecosystem or a species that already exists (because of its potential value).

In a report to the UNEP, Pearce *et al.* (1999) investigated the global economic value of biodiversity and the possible mechanisms by which such global economic values might be "captured," i.e. appropriated by the host nations in a cash, technology or resource form.

Two broad approaches to valuation have been mentioned: (a) direct approaches, for instance, direct use values include those derived from harvesting, subsistence and tourism, and (b) indirect approaches. The direct valuation approach elicits preferences by the use of survey and experimental techniques. Subjects are asked to state directly their strength of preference for a proposed change. Approaches which link some change in an environmental variable to a change in a marketed output yield a potentially large set of estimates.

An indirect approach may involve non-use valuation, i.e. willingness to pay (i.e. contingent valuation) to conserve, or accept in compensation for not exploiting a financial return, and thus maintain biodiversity. The option value method includes potential uses, direct or indirect, and insurance against the unknown.

## Modeling the economic value of medicinal plants

Pearce and Moran (1994) dealt with the economic valuation of medicinal plants in some detail. It can be derived in two ways. The first is based on existing use values which are directly related to commercial drugs and for traditional medicine, and the second relates to the *option value* of the plants, that is to say, the extent to which conservation is required to protect *future use values*. These, in turn, are critically dependent upon the basic plant sources upon which the future of medicinal drug research depends.

The economic valuation of plant-based drugs is dependent, in part, on the costs related to the exploration and acquisition of plant genetic material and the extent of demand by drug companies for that material. Companies or botanical gardens engaged in gathering plant specimens are paid under contracts or weight of material. Occasionally, they enter into royalty agreements with drug companies and are entitled to receive a share of the royalties, which may range from five to twenty percent, in the event of successful exploitation. However, as mentioned elsewhere, the traditional source of information concerning the medicinal properties of the plants — the indigenous tribes and local communities — has not been rewarded.

Royalties are more readily provided for plant material which is to be used in drugs that are close to being marketed. On the other hand, material that is yet to be screened for long-term development is likely to attract lower royalty agreements. Economic valuation can be based on several different factors:

(a) the actual market value of the plants when traded,
(b) the market value of the drugs of which they are the source material, and
(c) the value of the drugs in proportion to their life-saving properties.

It should be noted, however, that the financial reward realized by inventors often represents only a very small fraction of the value of the product to society because it is hard to appropriate all the qualities that are normally associated with a successful or popular product. These qualities include all the ecological and demographic variables that are associated with the habitat in the host country as well as the socio-political, economic, technological and other factors that are associated with drug development, which usually occurs in a developed country.

## The value of land for medicinal plants

The medicinal plant value of a hectare of "biodiversity land" can be estimated by using the following equation (Pearce and Moran, 1994):

$$V_{mp}(L) = p \times r \times a \times V_i(D).$$[¶]

## Probability of success ($p$)

From discussions with drug company experts, it has been estimated that the probability of any given plant species yielding a successful drug is between 1 in 10,000 and 1 in 1000. It has been estimated further that about 60,000 plant species are likely to become extinct in the next fifty years (Raven, 1988), which suggests that up to 60 species of medicinal significance could become extinct in the foreseeable future.

## The royalty ($r$)

As the existing royalty agreements range between five and twenty percent for drug development, a low figure of $r = 0.05$ is adopted for the purpose of this discussion.

---

[¶]$p$ = the probability that the biodiversity supported by a land (e.g. a hectare) will yield a successful plant-based drug $D$; $V_i(D)$ = value of the drug; $i$ = indicates one of two ways of estimating the value of the drug: the market price of the drug on the world market ($i = 1$), or the "shadow" value of the drug which is determined by the number of lives that the drug saves and the value of a statistical life ($i = 2$); $r$ = the royalty that could be commanded if the host country could capture all the royalty value attributable to natural capital (usually in the range five to twenty percent); $a$ = the coefficient of rent capture (one of the goals of the Rio Biodiversity Convention is to raise the value of $a$ which would benefit developing countries, thus providing an incentive for conserving biodiversity). If a host country could capture rent perfectly then $a = 1$. However, it could be as low as 0.1 in low income countries.

## The value of drugs ($V_i(D)$)

At the outset, two obvious points need to be emphasized. It would be erroneous to assume that drug prices truly reflect the cost of the plant source material as many other costs are included. Secondly, market prices often understate the true willingness to pay for drugs; there will be some who would be willing to pay more than the market price (and, of course, others who could barely afford to pay the market price of a drug).

According to Pearce and Moran (1994) only about 40 plant species accounted for the plant-based drug sales in the 1980s in the U.S. On the basis of prescription values alone, each species accounted for US\$290 million on average (US\$11.7 billion/40 = US\$290 million). The corresponding figure for 1990 for the U.S. was about US\$15.5 billion. On market-based figures, the *annual* loss to the U.S. alone was estimated to be about US\$8.8 billion, and to OECD countries about US\$25 billion.

Pearce and Moran (1994) arrived at an estimate of the value of a "representative" hectare of land. Their model can be written as follows:

$$V_{mp}(L) = \frac{N_R \times p \times r \times a}{H} \frac{V_i}{n}$$

where

$N_R$ = number of plant species at risk,
$n$ = number of drugs based on plant species,
$H$ = number of hectares of land likely to support medicinal plants, and
  with
$N_R$ = 60,000
$p$ = 1/10,000 to 1/1000
$r$ = 0.05
$a$ = 0.1 to 1.
$V/n$ = 0.39 to 7 billion US\$
$H$ = 1 billion hectares, the approximate area of tropical forest left in the world.

The resulting range of values is \$0.01–\$21 per hectare. If $a$ = 1 at all times, then the range is \$0.1–\$21 ha. The lower end of the range is negligible, but the upper end of the range would, for a discount rate of five percent and a long time horizon amount to be present value of about $450 ha (as estimated by Pearce and Moran (1994)).

## Valuing ecosystem services

It is of interest to examine briefly the philosophical basis of an intrinsic value from different points of view. There have been several discussions in recent years which were mainly concerned with the rational basis and the methods employed for valuing ecosystem services. For instance, the philosophical basis of value articulates the ethical basis or what constitutes the source of value and the empirical methods devise techniques for the measurement of that value which is based upon a particular philosophical notion.

## The utilitarian approach

The utilitarian approach recognizes value by employing direct as well as indirect methods. These may include a particular species which is an essential ingredient in our daily food supply, or other organisms that are a part of the food chain without being a direct component of our food supply. This approach involves both consumptive and non-consumptive use values. Numerous examples of all these types can be cited. Certain species may be consumed as food in some instances and the same species may serve as a link in the food chain in a different context. For instance, some bird species may give pleasure to bird watchers and naturalists, but they may also be consumed as food at times. Utilitarianism facilitates cost-benefit analysis in terms of gain or loss to different segments of human society. Ecosystems are complex social organizations in which the gain to society from one species may be intricately linked to that of another. Similarly, destruction or depletion of one species may simultaneously lower the value of other species as well. In addition, there are other dimensions to the problem of valuation. As pointed out by Goulder and Kennedy (1997),[**] there are certain

---

[**]See Daily, G. C., 1997.

ethical and moral dimensions that are beyond the cost-benefit approach. For instance, the benefits and costs may not be distributed equitably across various community groups, regions and generations. Indeed, this is already known to be the case in several instances such as those involving medicinal plants, fisheries and water resources. There are several ongoing controversies and litigations which involve these and many other resources.

Other views of valuation may not involve any utilitarian considerations. Fundamental rights, intrinsic rights and the ethical basis of all life are (or ought to be) involved in decision-making. Any monetary consideration of valuing ecosystem services is alien to the practitioners of several religions. Can one put a price on God's creation?

## Ecosystem services

Goulder and Kennedy (1997) classified ecosystem services into three broad categories: (a) the provision of production inputs, (b) the sustenance of plant animal life, and (c) the provision of non-use values.

## Characterization of some ecosystem services

Daily (1997)[††] attempted to value ecosystems and their constituent species in the context of the benefits derived from them, in the form of life-support goods and services. In addition to these fundamental material ingredients which sustain human communities, other direct and indirect benefits include aesthetic, historical, recreational and religious or other cultural values which have become essential ingredients of our civilization. A central feature of these benefits is the contribution of the genetic component which is the driving force behind the genetic diversity of ecosystems and their continuing evolution. Of course, there are several other kinds of influences as well which ultimately define the fate of natural ecosystems. There is also the constant threat of population expansion, which continues to destroy several natural ecosystems.

---

[††]Daily, G. C., *Nature's Services* (Island Press, Washington, D.C., 1997).

Technology, with its accompanying destructive impact of industrial and chemical pollution, has destroyed several precious natural ecosystems on land and water. Future policy regarding the protection and well-being of natural ecosystems should be guided by more efficient planning, allocation of effort and material resources. Daily (1997, pp. 370 and 371) outlined several broad questions which require research and investigation. These are summarized as the following:

(a) What is the enumeration of services provided by each ecosystem?
(b) What is the relationship between the quantity or quality of services and the condition of the ecosystem?
(c) What is the contribution of biodiversity to these services?
(d) What is the extent to which these services have been impaired?
(e) What is the nature and extent of the interdependence among these services?
(f) Are these services amenable to improvement, and, if so, to what extent, and what is the time-scale involved?
(g) To what extent can any future human technology replace existing ecosystem services?
(h) To what extent must we maintain the current geographical and ecological organization to ensure the continuity of the delivery of ecosystem services?

The question of valuation is an important one as it forms the foundation for future economic planning and allocation of sources. It is an undeniable fact that until recently most countries have not incorporated a comprehensive plan to protect natural ecosystems into their national planning programs. Some efforts have been made piecemeal in specific contexts, but now research is being directed increasingly towards the economic applications of natural ecosystems. Modern biotechnology and molecular biology are creating new opportunities for increasing commercialization and exploitation of the biodiversity of natural ecosystems. Increasing efficiency of modern transportation and the tourism industry have made it possible to appreciate the aesthetic and economic appeal of biodiversity all over the world. However, its importance to society is yet to be fully appreciated by political leaders and the public-at-large.

## Slowing down the loss of biodiversity

Economic incentives, agricultural reform and various conservation programs have been operating in several countries to slow down the rate of loss of biodiversity. Some of the conservation schemes include the allocation of land and acreage retirement programs, conservation payments for specific habitats and species, land easements and covenants, habitat zoning to restrict development and habitat replacement practices. These have been described in detail by Pearce and Moran (2000). One approach is to set priorities and choose which biodiversity should be saved at the expense of another. Should we save biodiversity for its own sake or should we be more concerned with the conservation of biological resources? Quite often, the two go hand-in-hand, as in the case of forest and wilderness conservation. Among the priority approaches are those that use the density of species in a given area, socio-economic (investment) approaches, and diversity based on phylogenetic patterns (Pearce and Moran, 2000). An economic approach to the conservation of biodiversity has been argued as a humane and anthropocentric viewpoint. This is not to deny the intrinsic value of preserving biodiversity. Other approaches to the conservation of biodiversity have been emphasized on the basis of aesthetic and moral values by others. However, it has been suggested that both the moral and intrinsic value approaches are not helpful in setting priorities for conservation.

A major aspect of the economic approach is its role in the maintenance of sufficient biodiversity to assure the continuity of eco-systems delivering services of value to the community. Species loss poses a grave threat to the resilience of ecological systems. Preservation of the *status quo* is less important or even undesirable in many situations.

An indirect threat to biodiversity is the high level of indebtedness in developing countries and the attempts of developed countries to appropriate "nature" or natural resources including biodiversity as a means to reduce the crushing debt burden. Such measures are not conducive to the long-term conservation of biodiversity because the creditors are invariably interested in immediate benefits rather than long-term planning. Needless to add, any claims of intellectual property rights by developing countries and their potential economic benefits are compromised under those strenuous circumstances. The exporting countries of the southern hemisphere have not been able to protect their natural resources adequately against the demands of the North. Indeed, it has often led to the environmental degradation of the South. For instance, only about three million hectares in the Netherlands are devoted to agriculture. However, another thirteen million hectares (an area equivalent to three times that of the Netherlands) in other countries is required to meet the demand for domestic consumption. Increased Dutch demand for tapioca resulted in a significant increase in Thailand's cassava production, which in turn contributed to extensive deforestation in northeast Thailand. Similar forest degradation was also noted in connection with the Dutch imports of soybean from Brazil and oil palm from Malaysia.

The problem persists because the environmental costs related to the production of exported products in developing countries are not being included in product prices. It should be collectively demanded by all developing countries.

**Pollinator Classes for Cultivated Food and Medicinal Plants of the World**[*]

| Pollen vector | No. of floral host species |
| --- | --- |
| Wind | 47 |
| All vertebrates | 155 |
| Birds | 52 |
| Bats | 103 |
| Thrips | 12 |
| Butterflies and moths | 35 |
| Flies | 179 |
| Beetles | 48 |
| All bees | 918 |
| Non-Apis bees | 796 |
| Honey bees | 122 |
| Wasps | 46 |
| Other insects | 66 |
| Unknown vectors | 549 |

[*]Modified after Nabhan and Buchmann, *The Forgotten Pollinators* (Island Press, Washington, D.C., 1997).

## Pollinators for the World's Wild Flowering Plants[*]

| Category | No. of taxa |
|---|---|
| Wind | 20,000 |
| Water | 150 |
| All insects | 289,166 |
|   Bees | 40,000 |
|   Hymenoptera (bees and wasps) | 43,295 |
|   Butterflies/moths | 19,310 |
|   Flies | 14,126 |
|   Beetles | 211,935 |
|   Thrips | 500 |
| All vertebrates | 1,221 |
|   Birds | 923 |
|   Bats | 165 |
|   Other mammals | 133 |
| **Total** | **640,924** |

[*]Modified from Nabhan and Buchmann (1997).

## Anthropocentric View of Biodiversity[*]

| Type of value | Features |
|---|---|
| Utilitarian | Approximately equals economic value |
| Naturalistic (and aesthetic) | Appreciation of nature and the beauty of wild species and habitat |
| Ecology (science) | Appreciation of complexity of nature and the scientific value of biodiversity |
| Symbolic | Use of nature to convey thoughts and ideas |
| Humanistic | Humanisation of wild species, domestication of animals and plants, bonding with nature |
| Moralistic | Kinship among all living beings, influence of Darwinian evolution |
| Negativistic | Negative values resulting from natural disasters |
| Dominionistic | Embedded concept of dominion over the challenge of nature |

[*]Modified after S. R. Kellert, *The Value of Life: Biological Diversity and Human Society* (Island Press, Washington, D.C., 1996).

## Estimated Medicinal Value of Plants[*]

| | |
|---|---|
| Mendelsohn & Balick (1995) | Net revenue to drug companies = US$2.8 to US$4.1 billion from rights of access to all tropical forests (about US$1.0 per hectare) |
| Simpson *et al.* (1995, 1996) | "Private" WTP of US$0.02 to US$2.29 per hectare of "hotspot" land; Max WTP of marginal species (US$9410) |
| Simpson & Craft (1996) | "Social" WTP of US$29 to US$2888 per hectare of "hotspot" land; Max WTP of marginal species (US$33,000) |
| Rausser & Small (1998) | "Private" WTP of US$0 to US$9177 per hectare of "hotspot" land |
| Farnsworth & Soejarto (1985) | US$298 million per plant derived drug (U.S.) US$2.4 million per year per single untested plant species (U.S.) |
| Principe (1991) | US$0.5 million peryear per untested plant spp. (OECD) |
| McAllister (1991) | US$9500 per untested tree species per year (Canada) |
| Principe (1991) | US$28.4 million per untested species per year (OECD) |
| Ruitenbeek (1989) | US$190 per untested species per year |
| Pearce & Purushothaman (1995) | US$743 to US$1.33 million per untested species per year (OECD) |
| Reid *et al.* (1993) | Up to US$4600 per untested species per year (annuitised at 5% over 20 years) |
| Artuso (1994) | Present value of US$866 per sample extract in terms of private WTP, US$9900 per extract in terms |

[*]Modified from Pearce *et al.* (1999); WTP = Willingness To Pay.

# The Role of Science in Developing Countries

At the outset, it is necessary to emphasize that the so-called developing countries are not all homogeneous with respect to their degree of development. They vary greatly in terms of their quality of scientific research and level of technological development. Any simplistic generalizations with an attempt to explain the role of science in all developing countries would be misleading. It would be no more accurate than an attempt to explain the growth of science and technology in Europe and North America in one sweeping generalization. For this reason, I am generally skeptical of the evaluation of the role of science in developing countries presented by Goldemberg,[1] although I agree with him on certain points.

In developing countries, the first priority for development ought to be the improvement of infrastructure, an obvious fact which Goldemberg failed to mention. Sophisticated science and technology cannot be expected to flourish in countries where a large fraction of the population lacks such basic amenities as safe drinking water and the minimum requirement of food and shelter. Poor communication and transport facilities are the rule in many developing countries. Energy is always in short supply. Power failures are a daily occurrence. Under these circumstances, is it any wonder that it is extremely difficult (if not impossible) to continue sophisticated research programs for any length of time?

Another serious problem, which Goldemberg failed to mention, is the rapid increase in population growth, which outstrips the prospective economic and social planning. In India, complex economic, socio-cultural and political factors have made it impossible to slow down population growth to any significant extent. Quite understandably, the population growth, in turn, has strained all available resources with the result that the basic amenities take precedence over the finer qualities of science and technology.

The latest technical information is not available in many developing countries. Without the latest technical information, it would be impossible to conduct a successful and competitive research program. I have been told quite often by the librarians of American universities that the cost of purchasing scientific books and journals has become prohibitively expensive. Surely, one can then imagine how much more difficult it is for a university in a developing country to maintain a high quality scientific library. This situation can be partially remedied if scientists in the developed nations would mail their discarded professional journals to a college or student in a developing country.

My knowledge of Indian science indicates that the Government of India encouraged and supported large research institutions which grew around an individual with proven distinction in a branch of science. Two examples are: the Atomic Energy Establishment in Bombay (now Mumbai), which came into existence entirely due to the efforts of Homi J. Bhabha, a brilliant nuclear physicist, and the Indian Statistical Institute in Calcutta which was founded and directed by P. C. Mahalanobis, also a brilliant physicist who switched to statistics. Both Bhabha and Mahalanobis were close friends of India's first Prime Minister, Jawaharlal Nehru. The key factor for their success are personal ambition, visionary leadership, friendship with the Prime Minister, professional distinction, and right man at the right time in the post-colonial India yearning for technological development. Goldemberg erred in not recognizing these intensely personal factors which led to the building of these large institutions. I doubt that competition with neighboring countries or imitating developed countries had anything to do with these developments. Two more recent examples: C. R. Rao's distinguished role in directing and shaping the Calcutta School of Statistics and assisting in national economic planning (which was first initiated by Mahalanobis), and M. S. Swaminathan's remarkable leadership in bringing forth the "green revolution," which has transformed India from a food-importing nation into an exporting one. In these instances too, there was no grandiose plan to imitate developed countries, but a desire and skill to meet domestic needs.

These successful scientists utilized, wherever possible, local talent and resources. Much emphasis on research and development in agriculture, fertilizers, engineering, infectious diseases, nutrition and oceanography (to name a few) was in direct response to domestic needs and economic concerns. Curiosity and a love for science are not incompatible with practical concerns.

There is, however, an irony which should be addressed in this context. In increasing numbers, immigrants from developing countries are directing and participating in the research and development programs of developed countries. The areas of activity include both basic research as well as technological applications. Large numbers of scientists, professors, engineers, physicians and others from developing countries are significantly contributing to the well-being of the industrialized world. Generations of individuals from developing countries are educating those in developed countries. Most of these immigrants have received their undergraduate education (and in some cases their graduate education) in developing countries. Goldemberg asked: "What is the role of science in developing countries?" Should we not ask: "What is the role of scientists from developing countries in developed countries?"[2]

## References

1. J. Goldemberg, *Science* **279**, 1140 (1998).
2. K. R. Dronamraju, *Science and Society: An Indo-American Perspective* (University Press of America, Inc., Lanham, MD, 1998).

# Indian Tradition in Science

India represents one of the oldest civilizations of great learning and knowledge. Imagine the time, thousands of years ago, when much of the world was occupied by barbarians and nomads. India was the center of great scholarship and sophisticated civilization. Indian mathematicians and astronomers reigned supreme. Great knowledge of botany and medicinal plants established the Indian pharmacopia, which formed the foundation for the Ayurvedic tradition in medicine.

Urban planning and public sanitation were among the most carefully laid out aspects of community living.

Unlike today's western science, scholarship of nature (as science was then called) was not divorced from the spiritual and aesthetic aspects of nature, but was nurtured as an integral part of all knowledge, a holistic view that is not found today. With this proud and noble background in Indian science, we must view present day India's science with some dismay and mixed feelings. And how do we account for this current state of affairs?

For almost eight hundred years, India had been under foreign domination. Although there were some isolated bright spots in mathematics and astronomy during those years, the state of former greatness never returned.

The period of occupation under British rule was not particularly congenial to the development of India's greatness in science.

This is perhaps a symptom of what V. S. Naipaul wrote in "India: The Wounded Civilization."

Only during the twentieth century did we witness the achievements of a few great scientists such as the physicist C. V. Raman, who was awarded India's first Nobel Prize in science in 1930 for his discovery of the light-scattering effect named after him. Other Indian pioneers in science at that time included the chemist Acharya Prafulla Chandra Ray, and the physicist turned botanist, Jagadish Chandra Bose. It should be mentioned that Rabindra Nath Tagore was India's first Nobel laureate (awarded in 1913) but his contribution was to literature, not science. We should also mention the great mathematician and a natural genius, Srinivasa Ramanujan (1897–1920), who made lasting contributions to mathematics (especially to number theory) although he never received any higher education in mathematics. From about 1930 onwards, the pace of science in India picked up, resulting in the election of several Indian scientists to the Fellowship of the Royal Society of London, but there were no other Nobel laureates until the 1960s when Har Gobind Khorana was awarded the prize for his contribution to the genetic code; and much later in 1982, S. Chandrasekhar (a nephew of C. V. Raman) received that honor for his great contribution to the theory of the origin of "black holes" and other aspects of astrophysics. Fortuitously, I came to know him well during his last years when he taught for many years at the University of Chicago.

It should be mentioned that the science of Khorana and Chandrasekhar, and more recently the Nobel Prize work of Amartya Sen in economics were all products of their careers in foreign countries.

Besides Nobel Prizes, there are other standards which can and should be used to judge the quality of scientific work. Judging by any standards, in recent years the large mass of Indian science has not produced more than a handful of capable scientists who are recognized worldwide, although India contains the third largest proportion of scientific manpower in the world. Furthermore, they have the advantage of being well versed in the use of the English language, which, for example, the Chinese and the Japanese do not. Indian graduates and scientists seem to perform well when they go abroad to pursue their scientific careers. However, inadequate resources, lack of capable leaders and a cumbersome administration seem to hamper the growth of science in India.

Here perhaps is a situation where NRIs can help to raise the standards of Indian science by contributing both resources and technical knowledge.

# The State of Science in India

I recently returned from India after participating in the Annual Meeting of the Indian Science Congress Association which was held in Chennai, and other scientific conferences in Chennai, Hyderabad, Mumbai and Delhi. I have participated in the Indian Science Congress over the last several years for a number of reasons. These include my genuine interest in the status of Indian science, my sentimental feelings as one who received earlier scientific training in India, and as one interested in promoting science and technology transfer to India. I am usually quite frustrated by the poor quality of science as well as the manner in which these conferences were conducted.

This is surprising because there are many well-qualified scientists in India who perform brilliantly when they work abroad. In searching for a solution, I am struck by several unfortunate and undesirable features: (a) Science in India is controlled by bureaucrats, not by scientists; (b) The leaders who frequent the science congresses are the same each year, playing musical chairs with the same group of older scientists exchanging positions of power and prestige at the top; (c) Medals are awarded to the same older scientists year after year. I noted that one older scientist was awarded yet another medal, adding to his large collection of medals and honorary doctorates. I was left wondering why there were not any young faces in the plenary sessions; (d) There is a lack of support for research by business and industry (almost all research is supported by the government); (e) A cumbersome bureaucracy persists which is made even more incompetent because of poor infrastructure (power failures and inadequate communication facilities just to name two of the daily problems which confront scientists and others); and (f) There is an inordinate fear of imported science or technology.

In spite of these problems, I have found bright young scientists working in a few selected laboratories. Science limps along and one hears of occasional suicides by young scientists. It is obvious that the fault lies at the top. No great leadership, neither political nor scientific, is apparent on the horizon. In spite of much lip service about the great Indian intellectual tradition, financial support for research and higher education is meager at best. Where progress can be achieved without any expensive facilities such as software development and statistical science, young Indians have performed very well in the international arena.

Under these circumstances, it is no wonder that droves of young Indian scientists leave for greener pastures abroad, not so much for financial gain, but to be able to work in an atmosphere of greater intellectual freedom and the opportunities to pursue their goals. In spite of the many frustrations and oppressive bureaucracy, I am impressed by the number of Indian scientists who continue to produce decent scientific work while often working under terrible conditions which no scientist in the western countries would accept.

But the cost-benefit ratio of scientific quality and productivity remains below par in several disciplines.

Perhaps science is a luxury in a country where many people cannot afford even the most meager amenities for food and shelter. Horrible streets, a total lack of public or private hygiene, frequent power failures, malnutrition and undernutrition, heavily polluted cities, and contaminated water supply are among the common problems.

Under these circumstances, it would seem ludicrous to hold any serious discussions about such topics as gene mapping and gene therapy.

# WSJ Critique of Sen's Views
# Seen as Wrong-Headed

On October 14, 1998 the Nobel Peace Prize for Economics was awarded to none other than Professor Amartya Sen, who is well known in India and throughout the developing world for his economic theories within an ethical and moral context.

This particular aspect, namely the ethical and moral consequences of economic policies, seems to annoy some experts in rich countries. Writing in the Wall Street Journal (dated October 15, 1998), columnist Robert L. Pollack bitterly attacked the selection of Professor Sen for the economics Nobel Prize.

In an article entitled, "The wrong economist won," Mr. Pollack wrote: "Throughout his long career and voluminous writings, he has done little, but give voice to the muddle-headed views of the establishment leftists who dominate his world of academics and non-governmental organizations... Mr. Sen has just been wrong.

When it comes to development economics, Professor Sen has focused on the importance of governments in promoting growth and bringing about more equitable resources. But it has become clear over the years that those countries that interfered least with their markets have done best... It would be nice to see the Committee recognize that, from time to time, by refusing to give an award, rather than default to someone of such debatable merit."

Generally speaking, western organizations and institutions tend to be hostile to ideas which present a more sympathetic view of the economic development of developing nations. In particular, they are usually opposed to any theory that is not consistent with the established market economics of western countries.

The Nobel Prize Committee and the Royal Swedish Academy of Sciences stated in their announcement that by combining tools from economy and philosophy, Professor Sen has restored an ethical dimension to the discussion of vital economic problems. His writings have examined the justice of economic policies, especially their impact on rich and poor. In fact, Prof. Sen held chairs in both the economics and philosophy departments at Harvard University. Currently, he holds the illustrious position of Master of Trinity College at Cambridge University in England — the first Indian to be appointed to that position.

## Childhood in India

Amartya Sen was born in Bengal in 1933. When he was nine years old, famine struck his region with great ferocity, killing three million people at a time when overall food production was not particularly low. Later, he earned his Ph.D. from Cambridge University, but returned to revisit his childhood trauma.

In a CNBC interview, he said: "I have always been concerned with the downside of economics, the miserable guys who end up hungry, unemployed, starving." It is not the economics of rich people, but the economics of poverty that has interested him throughout his life. In a news conference, he said: "When I say I'm an economist, people ask, 'What should I invest in?' But economics also deals with the downtrodden." One of his former colleagues said that he has brought a human face to economics. His best known book on the subject was "Poverty and Famines: An essay on Entitlement and Deprivation," published in 1981.

The Bengal famine made a big impression on nine-year old Amartya Sen. As an economist, he commented that the famine was inflation-induced. World War II brought a boom to the region. The presence of the British military resulted in an increase in wages and food prices in the cities. But in the countryside, the poor saw no increase in their income while facing higher food prices. Prof. Sen believes that famines never strike democracies, because elected governments are obliged to respond to their constituents' needs. On the other hand, autocrats and dictators feel no such obligation. British Colonial rule in India belongs to the latter category.

In awarding the Nobel Prize to Prof. Sen, the Nobel Foundation seems to recognize these historical injustices. Economics with a human face indeed. No wonder the Wall Street Journal feels differently. Most of the world's human populations live in poor countries. We wish Prof. Sen many more years of creative, healthy and happy life!

# Infrastructure Versus
# Technological Development

A revisionist view of India's development during its early post-independence years includes criticism of Jawarharlal Nehru, who, as the first Prime Minister, promoted heavy industries and massive large scale and glamorous projects, but ignored the development of a sound infrastructure.

To this day, after fifty-four years of independence, such elementary conveniences as telecommunications, highways and public transportation remain rather inferior in comparison to several other countries.

Outdated and defective infrastructure is a universal problem all over the world. Old and defective bridges, poorly maintained highways, inadequate supplies of clean water and air, and problems of communications, transportation, safe and nutritious food production, housing, safety in schools and quality of education are some of the obvious problems of infrastructure that are matters of concern in the U.S. However, these problems and others are much more acute in several developing countries because of lack of money, resources and trained personnel. The most obvious problem which confronts visitors to India and other developing countries is a lack of public hygiene, and garbage in the streets and other public places. It is everywhere.

Public toilets are lacking and when available, they are unclean and poorly maintained. Streets are freely used as public toilets. Drinking water is highly polluted.

It's no wonder that outbreaks of various infectious diseases are common. Infant and child mortality in developing countries continues to be very high in comparison to the situation in developed countries.

In spite of these problems of infrastructure, I see no great urgency on the part of the leaders to solve these pervasive problems. They face an obvious dilemma in utilizing their limited resources for the development of new technologies and modernization, as opposed to solving problems of infrastructure.

New projects are easily started because they are glamorous, sometimes necessary, and appeal to the public, so politicians like them.

However, the unglamorous task of maintaining old structures, buildings, and projects are almost totally ignored. Telephones and public transportation are absolute horrors! Consequently, except for a few shiny spots in the country, a visitor sees a depressing dilapidation in most public places. In such a prominent place as Connaught Place in New Delhi, people spit and urinate in public.

Not only are there no attempts to stop them, but no attempt is made to paint those buildings or refurbish the surroundings to make them less unpleasant; I would not even use the term attractive, because that would be too much to expect.

Of course, the air in Delhi is so polluted that it is hard to stay outdoors for long hours. Those who do, such as the traffic policemen, have to wear masks to protect their health.

The same situation prevails in several other large cities of the developing world, such as Mexico City and Bangkok.

Leaders of developing countries and their pseudo-middle class are chasing new technologies and consumer goods, trying to imitate the situation in developed countries. Yet, many fundamental problems of infrastructure are ignored.

Computers, electronics and various consumer goods are heavily favored while health research, higher education and basic sciences are ignored. Glamorous public impact projects are highly favored because of their political impact, but there is no glamour in maintaining infrastructure which is crumbling.

This prevailing attitude is short-sighted because it will lead to greater problems in the future. Not only will they have to rebuild a much deteriorated infrastructure, but they will also have to deal with the unfavorable consequences of a hastily developed technology.

# MISCELLANEOUS

# Controversial Sterilization of Women from the Developing World

A recent Associated Press report detailed how the drug, Quinacrine, is being freely distributed in poor nations through a network of doctors, nurses and midwives for the purpose of sterilizing women from the developing world. It is not approved by the FDA for use in the U.S. because of possible dangers including the causation of cancer. Its use is also opposed by most major family planning organizations, the World Health Organization (WHO), and many foreign governments. But that does not stop some doctors in North Carolina (who are tied to an organization opposed to immigration from developing countries into the U.S.) from supplying the drug at no cost to poor women in developing countries.

Quinacrine is inserted into the uterus in two doses (costing less than $2.00), a month apart. It prevents pregnancies by scarring the fallopian tubes. Side effects include abnormal menstrual bleeding, backaches, fever, abdominal pain and headaches. Doctors claim that women chose this method over surgical sterilization eleven-to-one when both were offered together in Vietnam.

Adrienne Germain, President of the International Women's Health Coalition in New York, said Quinacrine's low cost, portability and simplicity of administration make it readily open for abuse. This is especially true in poor countries which cannot afford to have such monitoring agencies as the FDA. She stated that such a drug should never be used anywhere in the world.

The Vietnamese government introduced a Quinacrine sterilization program in the 1980s, but stopped it in 1993 under pressure from the World Health Organization. The drug is distributed through the non-profit Center for Research on Population Security, funded in part by anti-immigrant organizations in the U.S. So far, more than 100,000 women in poor countries have been irreversibly sterilized by this method. The drug is freely distributed in the form of chemical pellets without a doctor's prescription. The procedure, if performed without anesthesia, can be painful. Some scientists believe that it can cause cancer in some women.

The organizations and individuals who are promoting this drug are clearly opposed to immigration from the developing world to the U.S. They are clearly racist in their sentiment because they are not opposing immigration from Europe, for instance.

Several years ago, the Nestlé Corporation promoted the sale of baby milk powder in African countries. However, it was not considered safe for sale in the U.S., and was not approved by the FDA. Later, its sale in Africa was halted because of pressure from the WHO and several developing countries. These issues raise fundamental ethical and moral questions. However, much exploitation of poorer countries continues in various forms. One such outrage is the patenting of traditional resources of the developing world such as neem tree products and basmati rice by European and American multinational companies. Current discussions in the WTO are directed at resolving these issues.

# My Tryst with Chandra

NASA's launching of their premier Chandra X-ray Observatory in the third week of July, 2000 reminded me of my meeting with Chandra five years before. The launch of the Chandra X-Ray Observatory is a proud moment in the brief Indian-American history in the U.S. Among those who witnessed the satellite launch was his widow, 88-year-old Lalitha Chandrasekhar.

The space shuttle Columbia that carried the Chandra satellite was commanded for the first time by a woman, Eileen Collins.

Dr. Subramanyam Chandrasekhar was born in Lahore (in the old British India) in 1910. When he was eight years old, Chandra's parents (Sita Balakrishnan and Chandrasekhara Subrahmanya Ayyar) moved to Madras (now Chennai). He studied physics at the Madras Presidency College, and later at Cambridge University in England, receiving a Ph.D. in astrophysics in 1933. He was one of the first scientists to combine the disciplines of physics and astronomy.

Early in his career, he demonstrated that there is an upper limit to the mass of a white dwarf star, now called the "Chandrasekhar Limit." A white dwarf is the last stage in the evolution of a star such as the Sun. When the nuclear energy source in the center of a star such as the Sun is exhausted, it collapses to form a white dwarf. This discovery is fundamental to much of modern astrophysics, since it shows that stars much more massive than the Sun must either explode or form black holes.

Chandra and his wife Lalitha immigrated to Chicago in 1937. He was on the faculty of the University of Chicago until his death in 1995.

Both he and his wife became U.S. citizens in 1953 and immediately took an active part in American politics, joining the Democratic party and supporting the presidential campaign of Adlai Stevenson.

Chandra was a popular professor who directed the research of over 50 Ph.D. students. It is said that Chandra used to commute twice-a-week to the University of Chicago's observatory in Wisconsin (in mid-west winters too) to teach a class. There were only two students in that class. Both received the Nobel prize, as did the professor, in later years.

Chandra's research covered all branches of theoretical astrophysics. He published over ten books, mostly in his field, but one discussed the relationship between art, science and poetry. He served as the editor of the Astrophysical Journal for 19 years while focusing his research on the structure and evolution of stars.

In 1994, I was developing my theory of "intellectual hybridization," which implies that great advances in science usually occur when the concepts and methods from two or more disciplines are combined. This is a gradual process of evolution of science, which is in contrast with the theory of discontinuous or abrupt revolutions that had been hypothesized by Thomas Kuhn earlier.

I corresponded with Chandra about my approach, and before long he invited me to visit him in Chicago. He sent me detailed directions on how to find his office on the campus of the University of Chicago.

It was a beautiful summer's day. We sat in his office and talked about science for two hours. He told me that he always arrived at his office at 10 a.m. every day. He was then eighty-four years old. He said his own work was an example of intellectual hybridization, combining quantum mechanics and the theory of relativity.

We talked about science in India, and about my mentor, J. B. S. Haldane. At one point, he showed me a book, and I pointed out that Haldane had a chapter in that book. He himself was busy then, writing a biography of his idol, Isaac Newton. That was his last book. It was post-humously published along with my book *Haldane's Daedalus Revisited*, and others by Oxford University Press in 1995.

After the meeting was over, he asked me politely if he could call a taxi cab to take me to the airport. He phoned for a cab, directed me to the University Faculty Club where the cab would be waiting, and returned to his office. The receptionist at the Faculty Club was helpful. A few minutes later, she announced: "Cab for Chandra." That's the way he was. In the world of astrophysics he was a giant, but to everyone else he was simply "Chandra."

We may justifiably celebrate the successful launch of the "Chandra Space X-ray Observatory." At 45-feet long, it was the largest satellite the shuttle had ever launched.

# Understanding Violence

I have written about the problems of violence in previous articles, but newspaper reports are once again filled with violence, both domestic and international. Recent school violence in Littleton, Colorado, shocked the entire world. The same page of the newspaper often carries a different headline, concerning the bombing of Yugoslavia by NATO forces. Purely political implications aside, the two reports have a common theme — killing of innocent civilians. I am amazed to see that the same television reporters and journalists, while regularly using harsh, condemning words when referring to the bombing and deaths in Yugoslavia, employ an entirely different technique — a soft and sad voice expressing deep sympathy and sorrow when reporting the deaths of school-aged children in Colorado.

This is, of course, quite understandable. Senseless killing of young students is shocking and terrible news. But, I see an entirely different picture; both reports give details of death and destruction; both reports give details of the violent deaths of innocent people. In Colorado, they involve several teenage students, and in Yugoslavia, many civilians of all ages, young and old.

## The causes of violence

I ask myself whether there is any connection between these two incidents. Our children are intelligent and perceptive. They grow up in an atmosphere of violence; watching violence on television, in the movies and in video games, in society, on the highways, and finally in the news. Children play with guns at an early age.

Who are their role models in resolving crises and social problems? What do they learn from watching the leaders and elders of our society? How are adults coping with domestic and foreign disputes? And what about the violence and crime in the streets? Are they not afraid to walk to school? Are they not afraid of other students and gang members? Fear begets fear.

They see how disputes are resolved by using guns, bombs and other means of violence. How often do they see alternate models and peaceful means of conflict resolution? Is it any wonder that someone runs amuck occasionally, killing a lot of others? I am surprised that it doesn't happen more often!

There is much truth in the old adage that "no man is an island." We are all influenced by what goes on in our society. The distinguished psychobiologist, Dr. Robert M. Yerkes, once said: "One chimpanzee is not a chimpanzee at all."

Scientists at the Yerkes Laboratory at Emory University in Atlanta conduct research on the social behavior of chimpanzees and other animals, which can lead to a better understanding of human behavior and human societal problems. Animals rarely kill others of their own group or species. Animal aggression is mainly directed at killing other species for food or to survive in the face of a clear threat.

The famous behavioral scientist and Nobel Prize winner, Konrad Lorenz, stated that there is no "sport" killing in animals. I recommend that everyone read Lorenz's book, "On Aggression," which is available in public libraries.

Humans are the only species who indulge in killing as a "sport." Of course, the same guns that are used for sport can also kill fellow humans.

More efficient guns that are useful for rapidly killing many enemies in warfare are also being used in killing school children. Gun control is the subject of much debate in our society. The politics of gun control are beyond the scope of this article. But, I believe that easy access to guns and bombs is really asking for trouble!

Social influences shape our emotional and physical development. But each one of us reacts differently in any given situation.

Obviously, some individuals react with great violence. Other may "turn the other cheek," but they are rare in our society today. It is the combination of emotionally disturbed people and the easy availability of guns together that leads to social violence and deaths.

Unfortunately, a radical solution is beyond our capacity. The hero in many mythologies, fiction, movies and video games is the one who kills the most people. Our society is not kind to "wimps." This is a much misunderstood concept. Mahatma Gandhi showed that it takes great strength and courage to practice non-violence. Gandhi provided an alternative role model for our children.

Violence is increasing all over the world. This is mainly due to overpopulation and a competitive struggle for land, food, water and other resources. However, violence and deaths are more prevalent in the U.S. because of the easy availability of guns and other destructive weapons. Fortunately, they are scarce in many other countries, especially in developing countries, where most of humanity lives. Konrad Lorenz observed that our society and social order depends entirely on our moral responsibility.

# The Hunger Project

Many of us are not familiar with the Hunger Project. Professor Amartya Sen's Nobel Prize has once again focused attention on starvation and famines. The Hunger Project is a global organization of loosely knit voluntary community groups based in countries all over the world, working with the singular goal of eradicating hunger in the world. I am proud to say that the Chairman of the Global Board of Directors is none other than the distinguished scientist from Madras (Chennai), Professor M. S. Swaminathan, who is well known in India as the father of the "Green Revolution." It was Dr. Swaminathan who turned India from a food-importing nation into an exporting one, for which he was awarded the honor of Padma Bhushan. The President of the Hunger Project is Ms. Joan Holmes. I was honored to be present at the annual meeting and banquet in New York in 1998. The mistress of ceremonies was the well known actress Ms. Valerie Harper. She has devoted her life to helping the Hunger Project. In her Presidential Address, Ms. Holmes said: "Your Hunger Project today is catalyzing a phenomenon. It is unleashing the creativity and productivity of hundreds of thousands of the world's poorest people. This process of empowerment is contagious — it is spreading. To mention only two examples: (1) in Bihar, India, the 60 women's fishing cooperatives established last year are now reaching out to create 800 more cooperatives, and (b) in Bolivia, our partnership with Radio ACLO is empowering more than 500,000 villagers in the Andes with literacy instruction and critical information on health, agriculture, women's empowerment, and participation with local government, etc.

These breakthroughs are occurring in the space of love and commitment of a growing global team of investors."

## Feminization of poverty

What impressed me in Ms. Holmes' address was her emphasis on the increasing "feminization of poverty." Hunger and poverty are closely related. Most of the world's poor are women. Seventy percent of the world's illiterates are women. Seventy-five percent of the world's refugees are women. There is more hunger in countries which are noted for lack of women's progress. In many developing countries, as children, women are under-nourished, work backbreaking 16-hour days, and are often in poor health. They are often undervalued, fed less, given inadequate healthcare and denied education. As teenagers, they are forced into early marriage (often with the crushing burden of dowry). One woman every minute dies each year from pregnancy-related complications. Most of these deaths are preventable. The low status of women has far-reaching and profound consequences. It affects children quite adversely. For instance, high rates of child malnutrition are caused by the low status of women. A landmark 1996 study by UNICEF revealed that South Asia has the highest rates of malnutrition in the world.

Even under inhumane conditions, women in the developing world are some of the most productive members in the world community. Women account for seventy to eighty percent of household food production in Africa, sixty-five percent in Asia, and forty-five percent in Latin America. Seventy percent of Africa's farmers are women, yet only seven percent of Africa's agricultural extension services are devoted to women farmers. Laws prevent women from becoming landowners in many developing countries. Adding to these problems, frequent civil wars have caused a great deal of hardship in many parts of the world, namely in Nigeria, Ethiopia, Cambodia and Vietnam in recent years. Once again, most of the victims are women and children. Poor countries spend their meager resources on costly armaments, not food production and healthcare.

The Hunger Project is a great humanitarian project which deserves support from all of us. Many of us are very fortunate in being able to afford a comfortable lifestyle, more than adequate nutrition and housing, and higher education. But most of the world's poor in Asia and Africa live in shocking poverty. Starvation and hunger are frequent occurrences. Yet, how many of us make any serious attempt to help those in great need? And how often? A sustaining long-term help is urgently needed. That is the goal and function of the Hunger Project.

172

## Further Progress Made for the
## Hunger Project

In 1999, the 22nd annual meeting of the Global Board of Directors of the Hunger Project was held in Chennai, India. It was chaired by Dr. M. S. Swaminathan, the distinguished agricultural scientist who has spent much of his life improving the world's food production. Actress Valerie Harper has contributed much of her time and effort to help the Hunger Project. Most of us do not give much thought to hunger expect for a brief period before lunch or dinner time.

However, the much larger issue of chronic persistent hunger, is that it is a silent killer that takes the lives of 24,000 people every day. Three quarters of these victims are children under the age of five years. According to the recent World Food Summit, 840 million people or one-seventh of the world's population live in the condition of chronic persistent hunger. The majority of hungry people live in South Asia and Sub-Saharan Africa.

It is an ironic tragedy that those who suffer from chronic persistent hunger are in that condition not because there is not enough food in the world, but that they lack opportunity to earn enough money to buy adequate food. In some countries, no matter how hard people are willing to work, they cannot earn more than US$1 per day.

So, the problem of chronic, persistent hunger cannot be alleviated by mere technical or financial aid. It requires sustained and comprehensive social reorganization. The Hunger Project recognizes the urgent need to reorient the very framework of thinking with respect to our conception of what hunger really is, and how we can go about correcting this chronic social problem.

One method of measuring hunger is by comparing infant mortality rates (IMR), the number of children who die before their first birthday. When IMR falls below 50 in a population, it is generally agreed that hunger has crossed a critical threshold, ceasing to be a society-wide issue in that population.

Much progress has been made during the past fifty years to eradicate world hunger than in the entire period of the last two millennia. Much of this progress was due to the contribution of the Hunger Project since it began in 1977.

IMR rates* have fallen below 50 in Latin America since 1977. However, in spite of some progress so far achieved, they continue to remain well above 50 in South Asia and Sub-Saharan Africa. On the other hand, in Europe and North America, they have remained below 50 during the last several decades. The world average IMR has come down from 103 in 1977 to 59 in 1997.

Dramatic changes that have come about in the world have contributed to the progress so far achieved in the world hunger situation. These are:

(a) as emphasized by the Nobel Laureate Amartya Sen, famines do not occur in democracies, and democracies with free press have flourished;

(b) the end of apartheid, and the end of the Cold War have ended conflicts that have devastated developing countries;

(c) there has been a significant social reorganization that has empowered women and minorities, village self-government groups, farmers' cooperatives, women's groups, youth groups, professional and sub-professional groups, and flourishing opposition political parties, to name a few;

(d) women's participation, child safety, environmental concerns and population growth have moved to center stage in many societies; and

(e) the absence of a global war since 1945.

---

*IMR rate is measured as deaths per 1000 births.

In spite of growing optimism, challenges will remain. Regional conflicts are a constant occurrence, resulting in large numbers of refugees. As usual, women and children are the main casualties. Conflicts disrupt agriculture and food production, creating famine and disease. Embargoes of various kinds raise IMR rates. Economic catastrophes, plummeting stock markets, and evolving pests pose a constant threat to food production, human health and the environment. The Hunger Project will be needed for a long time to come.

# Recipients of the National Medal of Science and Technology Announced

When the White House announced the names of distinguished individuals who were selected to receive the 1999 National Medal of Science and the National Medal of Technology, in the science category, there were twelve medals which were classified as follows: 1-molecular biology, 1-cell biology, 2-mathematics, 2-chemistry, 1-physics, 1-atmospheric science, 1-environmental science, 1-economics, 1-astronomy, and 1-computer science. Only five technology medals were announced: 1-medical technology, 1-biotechnology, 2-computers, and 1 for wireless technology. The message is clear. Basic sciences need to be promoted and emphasized. This is especially true of mathematics and science in the U.S. where students have not compared favorably with the graduates of other countries. Mathematics and computer technology are now receiving more attention than in previous years and the quality of school education in general has played a prominent role in campaign politics.

Bill Clinton showed a keen interest in biotechnology, especially genetic engineering and the genome project. Much of the research in this area is concentrated in private industry, especially pharmaceutical and large agribusiness companies. Some universities (such as Washington University in St. Louis), and U.S. Government laboratories, (such as those of the National Institutes of Health and the U.S. Department of Agriculture) are also active in this enterprise.

## Proclamation: National Biotechnology Month

Clinton's emphasis on research in biotechnology was clear in his presidential proclamation for National Biotechnology Month, January, 2000. The proclamation declared that great potential benefits of biotechnology-derived innovations are recognized by the administration. The biotechnology industry has stimulated the growth of several businesses, creating employment for 150,000 people, and investing ten billion dollars per year on research and development. New cures for diseases such as Parkinson's, Alzheimer's, diabetes, AIDS, heart disease and cancer are being developed with the aid of new biotechnology. Small business innovation research and intellectual property protection are recognized as important components of the new biotechnology industry.

The importance of biotechnology to agriculture is well recognized. Bioremediation technologies are improving our environments by removing toxic substances from contaminated soil and ground water. Genetic engineering and other DNA-based methods are reducing our dependence on pesticides. Manufacturing processes based on biotechnology enable us to produce paper and chemicals, involving less energy, less pollution and less waste. DNA fingerprinting technology is helping to exonerate those who are falsely accused, but bring those who are guilty to justice.

Without question the importance of the National Medals in Science and Technology, awarded by the President each year, and their emphasis on basic sciences, cannot be understated.[*]

---

[*]Dr. Neal Lane was Science Advisor to President Clinton.

# International Genetics Congress in China

In August 1998, I was in Beijing, China, at the invitation of the Chinese Academy of Sciences, to participate in the Eighteenth International Congress of Genetics, which was held at the Beijing International Convention Center. There were two thousand delegates from over fifty countries. I accepted an invitation from the Chinese to chair two symposia: one on the subject of "Environmental carcinogens and prevention of cancer" and another on "Deleterious mutations." Having read a great deal about the political and social situation in China in the American press, I was curious to observe and gather my own first-hand impressions.

First, I was happy to note that our scientific sessions started on time. Second, the Chinese scientists worked very hard to make it a success. Third, my hosts were most polite and hospitable, more so than at other scientific meetings that I had attended in Europe and North America. Fourth, our discussions took place in an atmosphere of cordiality and free expression. If there was any government interference in the conference, it was not obvious to me. The only government official that I recall was a Vice-Mayor of Beijing who welcomed us during the inaugural session. Fifth, the most important point of all, I enjoyed peace and quiet for a week without being bombarded by commercials and billboards or loud music. There were no reports of daily murders, guns and drugs in schools, or misbehavior of officials in Washington and elsewhere. It was a great relief not to be confronted by reports of the sexual exploits of Monica Lewinsky during those days. My colleagues from abroad and I felt safe walking in Beijing at night. The streets were spotlessly clean. We were told that while such international conferences are a common practice in western countries, they are regarded by the Chinese as an exceptional opportunity to show the best of Chinese science and society.

Now for a different perspective; some foreign colleagues remarked that the apparent calm we had witnessed in the streets and elsewhere is due to the highly repressive manner in which the Chinese government deals with any dissent in their society. Perhaps that was true, nevertheless it was a nice change from the noisy commercial-ridden western society, at least for a few days.

Biodiversity is protected in China because there are very harsh laws against exploitation and destruction of rare plant species, especially medicinal plant species. A systematic conservation program specifies the manner of harvesting, stage of growth of plants at the time of harvesting, parts to be harvested and the amounts harvested. Gene banks have been established to conserve germplasm of medicinal plants and other species of value. The Institute of Medicinal Plant Development (IMPIAD) is a WHO (World Health Organization) collaborating center of Traditional Medicine in Beijing which specializes in the research of medicinal plants under the auspices of the Chinese Academy of Sciences (CAMS).

The program of the Genetic Congress showed that the Chinese were not familiar with the latest advances in the field of genetics, but they were willing to listen and learn quickly. The President of the Congress, C. C. Tan, studied with the famous Drosophila geneticist Theodosius Dobzhansky, at Columbia University in New York, many years ago. Although in poor health, he managed to attend some of the plenary sessions of the Congress and mingled with the local students, who admired him greatly.

We enjoyed visiting the historical and tourist sites; Tianamen Square in the heart of Beijing where several historic buildings including a mausoleum for Chairman Mao and the Great Hall of the People are located. A short drive from Beijing took us to the Great Wall of China, which is about 6700 kilometers long. As it stands today it is a Ming creation, although successive earlier dynasties added to the wall from time to time. It is a solid well-built wall which varies in width from ten to twenty feet or even more in certain places. These were fortifications where soldiers used to keep watch, protecting the city and the surrounding area against the barbaric northern nomads. In Beijing, we saw the Lama Buddhist temple which was built in 1694, The Palace Museum and the Forbidden City (home of 24 emperors, constructed during the years 1406 to 1420 A.D., covering 74 hectares); and the Beijing Zoo with the famous pandas. A visit to Peking Opera was a unique experience. It combined comedy with fine dance and shrill music as well as great balancing feats.

During the conference, we discussed a controversial Chinese law which prevented childbearing, when genetic defects might be involved, through forced sterilization or long-term contraception. What makes the situation even more serious is the fact that Chinese couples are only allowed to have one child. Each community is watched over by an inspector who makes sure that this rule is not violated by anyone. Secretly giving birth to a baby will result in a heavy fine and a forfeiture of all personal property including home, furniture, clothes, jewelry and all other belongings. Many western scientists boycotted that Genetic Congress as a protest against this eugenic law. Others, like myself, went there because we preferred an open dialogue with the Chinese geneticists. I believe we made the right decision, but I left with one regret. Several younger (Chinese) scientists approached me to ask for my help in finding a research position in Europe or America (as happens to me often during my annual visits to India also). But what they were seeking is beyond my resources.

# Why Should We Be Interested in the Census?

Then U.S. President Bill Clinton visited Houston in 2000 with the stated purpose of addressing a Hispanic group about the need for collecting accurate census data. He briefly reviewed the historic background of the census in the U.S.

Census dates play a vital role in the planning and execution of various social programs, allocation of federal funds, and most importantly, in recognizing the extent of the ethnic diversity of the population of a country. The census forms the basis for building more highways, more hospitals, more child care centers, more post offices and other communication centers, to place more law enforcement personnel, and to fund a great number of social and community programs which are vital to the existence of a successful and civilized community. In addition, it affects our very democratic process, recognizing all of the groups and subgroups within a voting population.

The census is conducted once every ten years. The taking of the census has been mandated by the Constitution from the time the nation was founded. During the very first census, Thomas Jefferson, who was then Secretary of State, actually sent federal marshals out on horseback to count heads. For almost two centuries, this method of counting was followed, employing large number of field workers to count every individual in every household all across the nation. But as the population expanded and become more mobile, this method became inefficient and impractical. Numerous mistakes were made. Some groups were either underestimated or entirely missed. These mistakes, in turn, led to political problems because many communities felt that they were not receiving adequate federal funds or social amenities to which they were entitled.

Counting by mail was implemented in 1970. For the last three decades, this method has been used. However, millions of people did not return their 1990 census forms and the resulting census data were not at all accurate. Eight million Americans living in inner cities and remote rural areas were not counted. On the other hand, four million house-owners were counted twice! For instance, the number of people missed in Los Angeles alone will fill the city of Tallahassee, the capital of Florida. The census missed almost half a million people in the State of Texas, of which about 70,000 lived in Houston. More than half of those missed were children, which meant that the needs of children and related social programs were underestimated.

## Statistical sampling

In order to remedy these mistakes, the National Academy of Sciences recommended using a statistical sampling method with supplementary head counts in selected areas. The degree of error could then be reduced to less than a tenth of one percent. In a population census of 300 million people, we would then miss only 300,000, as compared to eight million in the 1990 census. It would also be cost-effective. Sampling methods have been used for many years in other fields of work; for instance, in the estimation of crop yields, household income, or in planning public health programs for infectious diseases, etc. Statistical sampling methods have been developed by some outstanding statisticians such as Sir Ronald A. Fisher in England, Professor P. C. Mahalanobis, the founder of the Indian Statistical Institute in Calcutta, and Professor C. R. Rao, who is now the elderly, distinguished Professor of Statistics in Pennsylvania State University. Indeed, statistical sampling has been pioneered in India to a very great extent and has played an important role in India's successive five-year plans. In a population census which utilizes the statistical sampling method, various ethnic minorities and all of the sub-groups in the country will be included. No group will be missed. As a small minority in this country, Indo-Americans will not be under-estimated or omitted, but will be represented accurately in the future census.

# Index